VOICES FOR EVOLUTION

VOICES FOR EVOLUTION

Revised Edition Edited by Molleen Matsumura

Introduction by Isaac Asimov

THE NATIONAL CENTER FOR SCIENCE EDUCATION, INC.
Berkeley, CA

Library of Congress 95-74815

Voices for evolution

ISBN 0-939873-53-2

Published by The National Center for Science Education, Inc.
        P.O. Box 9477, Berkeley, California 94709.
Printed and bound in the United States.

TABLE OF CONTENTS

PART THREE: RELIGIOUS ORGANIZATIONS

PART FOUR: EDUCATIONAL ORGANIZATIONS

PART FIVE: CIVIL LIBERTIES ORGANIZATIONS

FOREWORD TO THE FIRST EDITION

This book is the unique conception of Dr. Kenneth Saladin, Georgia College, Milledgeville. It was his brain child to gather together resolutions, statements, and position papers from organizations — scientific, educational, and religious/philosophical — which presented the views of groups of people on the creation/evolution controversy. He did all the groundwork and set the collection well on its way before yielding it to me to edit when he was pressed by other commitments.

There are two apparent exceptions to our editorial policy of offering only statements from organizations: remarks from the Episcopal Bishop of Birmingham and from Pope John Paul II. We elasticized our policy here because each man spoke in his official capacity as representative of members of his organization.

Voices is a project of the National Center for Science Education, an umbrella group set up in 1983 to support and coordinate activities of local, autonomous Committees of Correspondence. Most CCs were founded, beginning in 1981, by Stanley Weinberg, retired master biology teacher and author of biology textbooks. Weinberg understood that creationists, regardless of how their court cases are decided, work effectively at the grassroots level and should be dealt with there. From the first two committees, in Iowa and in New York, there are now 50 in as many states and five in Canada. Explains Weinberg:

The creation/evolution controversy is not an intellectual or scientific dispute, nor is it a conflict between science and religion. Basically, it is a contest over control of educational policy.

The short-term, immediate goal of NCSE and the CCs is to keep "scientific" creationism from being taught as legitimate science in public schools. The long-term goal is to improve science teaching, and the public understanding of science. Evolution — the fundamental organizing principle of biology — has been taught so little and so poorly that creation "science" has made inroads the scientific community wouldn't have believed possible.

It must be emphasized that no scientist disputes the right of fundamentalist Christians to believe that Genesis is a history and science textbook. The only difficulty arises when they seek to bring their sectarian religious faith into public school biology classes as legitimate science. The various statements here, from their various perspectives, ringingly declare, again and again, like variations on one mighty theme, that religion and science, properly viewed, can enhance and complement each other, but that they are different disciplines which deal in different ways and for different reasons with different spheres of human discovery. To blur that distinction weakens both.

Among the many, many persons who made this book possible, I want to give special thanks to Dr. Don Huffman, Central College, Pella, Iowa, who undertook the formidable task of getting permissions to use copyrighted material. Special thanks, too, to Dr. John Patterson, Iowa State University, Ames, and his assistant Gee Ju Moon, a genius with computers, who prepared the manuscripts in their many versions. Jodi Griffith designed the cover, and Liz Hughes the book layout. Thanks to friends across the country who read about the project and believed in it and contributed helpful suggestions and statements from their organizations.

All concerned hope that the book will be valuable, even invaluable, to biology teachers, boards of education, school superintendents, and librarians when they must respond appropriately to creationist demands.

Betty McCollister
Iowa Committee of Correspondence

FOREWORD TO THE REVISED EDITION

The clamor for teaching creationism has grown louder in the years since *Voices for Evolution* was first published. It is as if the call for creationism was once a faint rumble of thunder on the horizon, and now, in more and more communities, the lightning is striking. That's why a new edition is needed — not just to update the text, but because nearly every volume we printed has been distributed and put to use. Fortunately, as the threat has grown more obvious, the strength of the response has grown accordingly, and that is why we were able to add more voices — including legal opinions and a group of civil liberties organizations, as well as additional contributions to the sections for educational, scientific, and religious organizations.

Preparing this revised edition of *Voices for Evolution,* like working at the National Center for Science Education, has been a constant source of that special brand of pride and inspiration that comes from working with a skillful and dedicated team. With the first edition, Kenneth S. Saladin and Betty McCollister provided a strong foundation to build on. Countless friends and supporters of NCSE suggested possible contributors to this new edition, and the contributors themselves have been most generous with their time and eloquence.

The voices in this book, even though each one represents an important group of concerned people, are not the entire chorus. They are like the instrument that sounds the pitch so all the singers can work together. The other voices for evolution are teachers and parents, school administrators and concerned citizens and scientists — thousands of people across the country who work to defend evolution education in their own communities. This book is for everyone who wants to join the chorus.

Some years ago, a book appeared with the provocative title, "Steal This Book." I'd like to suggest something even more provocative: USE THIS BOOK! Use it when you're answering a creationist letter to the editor of your local newspaper, quote it to school boards and textbook adoption committees when urging them to adopt texts that teach evolution, donate it to a school library as a way of showing the staff your support for good science education. If you are fortunate enough to live in a community where good science is appreciated — and taught in the schools — give the book to a friend who lives in a community where voices for evolution need to be heard.

But *Voices for Evolution* is more than a tool for accomplishing a task — even an all-important task like defending truth in science and science in education. It is also worthwhile reading. Like the music that a chorus comes together to sing, the content of the book is inspiring. The entries are informative and perceptive. Use the book, and use it well, but please enjoy it, too.

Molleen Matsumura, Network Project Director
National Center for Science Education
August, 1995

ACKNOWLEDGMENTS

"Scientific" creationists claim that theirs is a legitimate scientific endeavor, and that there is "abundant scientific evidence" that the world and its life forms came about exactly as described if Genesis is interpreted literally. The *Voices for Evolution* statements from scientists refute this claim of scientific evidence for what Craig Nelson calls "quick creation."

Another claim advanced by "scientific" creationists is that those who accept the idea that evolution took place are anti-religious. The statements in this volume from religious organizations make that idea untenable, a position with which the statements from educational groups agree. These two groups, along with scientific organizations, find that evolution is scientific, creationism is not, and that the Biblical literalist view is not the only view acceptable to religious people. As this book documents, mainstream Judaism and the major Christian denominations, both Protestant and Catholic, have no difficulty accommodating evolution to their religious perspectives.

We hope *Voices for Evolution* will assist in spreading this important message to members of the public and those responsible for the decisions which shape our children's educations.

The National Center for Science Education is funded by subscriptions, donations, and grants from a number of private foundations. The first edition of *Voices* was funded primarily by donations from the Deer Creek Foundation, the Carnegie Foundation of New York, the Richard Lounsbery Foundation, and a foundation which wishes to remain anonymous, as well as donations from members; the Esther A. and Joseph Klingenstein Foundation provided additional funding for the second edition. We wish to thank all of these individuals and organizations for making the work possible.

Eugenie C. Scott, Ph.D., Executive Director
National Center for Science Education

INTRODUCTION:
Science versus Creationism

There is a belief system called "creationism" that calls itself "scientific creationism" in an attempt to make itself gain legitimacy. It is important to understand why this use of a respectable and admired adjective is, in this case, nothing but a disgraceful imposture.

Science is a process of thought, a way of looking at the Universe. It consists of the gathering of observations which can be confirmed by others using other instruments at other times in other ways. From these confirmed observations, consequences and conclusions can be reasoned out by logical methods generally agreed upon. These consequences and conclusions are tentative and can be argued over by different people in the field and modified or changed altogether if additional, or more subtle, observations are made. There is no belief held in advance of such observations and conclusions except that observations can be made, that consequences and conclusions can be reasoned out, and that the Universe can, at lease to a degree, be made comprehensible in this fashion. (If these assumptions are not true, then there is no way of using the mind at all.)

Creationism, on the other hand, begins with a strong and unshakable faith to the effect that all the words of the Bible are literally true and cannot be wrong. The function of observation and logic is then confined to the confirmation of the literal meaning of the words of the Bible. Any observation, or any course of logic, which seems to argue against those words must then be false and must be dismissed. Any conclusions of science that seem to argue against those words must also be false and must be dismissed. To find some excuse to do this without seeming entirely arbitrary, creationists do not hesitate to distort scientific findings, to misquote scientists, and to play upon the emotions and prejudices of their unsophisticated followers. Whatever creationism is, then, it is not scientific.

Science works through the organization of many observations that may in themselves seem to have no interconnection. Such organization is called a "theory" that demonstrates interconnection, gives meaning to the observations and, very often, predicts as-yet-unmade observations. Such a theory is rendered the more valid as more and more scientists make observations that fit the theory. However valid such a theory may seem, it is always subject to modification and further generalization, of course. Such modified and generalized theories are always stronger and seem still more valid because of what has been introduced. The theory of evolution is extremely strong, and what modifications have been introduced since Darwin's time have but made it ever stronger until now it is the very backbone of biology, which would make no sense without it. (And mind you, biology consisted of a miscellaneous set of observations that made no real sense before the theory of evolution was introduced.)

Creationists, on the other hand, have no theories, since they accept as true only the literal words of the Bible, which represent miscellaneous statements that do not support, and often contradict, each other. Their method of dismissing a scientific theory such as that of evolution is, in the main, to define a theory, arbitrarily and ignorantly, as "a guess." There is no trace of anything scientific in creationism, therefore.

Science depends upon the decisions of the intellectual marketplace. All its observations, all its conclusions, all its theories, are openly published and are studied and argued over. There are controversies and disputes that are sometimes unresolved for long periods of time. There are even (since scientists are human) observations made, sometimes, that are false or conclusions that are unjustified. These are sooner or later discovered by other scientists, since it is hard, or even impossible, to maintain for long an imposture in the face of the scientific system of open investigation.

Creationism, on the other hand, cannot endure the intellectual marketplace, since it will not allow its basis to be questioned. The literal words of the Bible are asserted as true to begin with; how, then, can there be any questions, any arguments, any controversy? This is, of course, unscientific gibberish. In order to fight this inevitable dismissal, creationism calls on the power of the state to force it to be taught as science. This would make it possible for politicians, under pressure of their own ignorance, or the lack of sophistication of their constituents, to take it upon themselves to define what is science. If politicians can do this, they can define whatever they choose, however they choose, and our every liberty is in jeopardy. By demanding political action, creationism turns itself into a political force, and is less than ever a scientific one. — And it makes of itself a great danger.

Isaac Asimov

PART ONE

Legal Background

BACKGROUND:
Six Significant Court Decisions Regarding Evolution/Creation Issues

1. In 1968, in *Epperson v. Arkansas*, the United States Supreme Court invalidated an Arkansas statute that prohibited the teaching of evolution. The Court held the statute unconstitutional on grounds that the First Amendment to the U.S. Constitution does not permit a state to require that teaching and learning must be tailored to the principles or prohibitions of any particular religious sect or doctrine. (*Edwards v. Aguillard* (1987) 482 U.S. 578)

2. In 1981, in *Segraves v. State of California* the Court found that the California State Board of Education's *Science Framework,* as written and as qualified by its anti-dogmatism policy, gave sufficient accommodation to the views of Segraves, contrary to his contention that class discussion of evolution prohibited his and his children's free exercise of religion. The anti-dogmatism policy provided that class discussions of origins should emphasize that scientific explanations focus on "how," not "ultimate cause," and that any speculative statements concerning origins, both in texts and in classes, should be presented conditionally, not dogmatically. The court's ruling also directed the Board of Education to widely disseminate the policy, which in 1989 was expanded to cover all areas of science, not just those concerning issues of origins. (*Segraves v. California* (1981) Sacramento Superior Court #278978)

3. In 1982, in *McLean v. Arkansas Board of Education,* a federal court held that a "balanced treatment" statute violated the Establishment Clause of the U.S. Constitution. The Arkansas statute required public schools to give balanced treatment to "creation-science" and "evolution-science." In a decision that gave a detailed definition of the term "science," the court declared that "creation science" is not in fact a science. The court also found that the statute did not have a secular purpose, noting that the statute used language peculiar to creationist literature in emphasizing origins of life as an aspect of the theory of evolution. While the subject of life's origins is within the province of biology, the scientific

community does not consider the subject as part of evolutionary theory, which assumes the existence of life and is directed to an explanation of how life evolved after it originated. The theory of evolution does not presuppose either the absence or the presence of a creator. (*McLean v. Arkansas Board of Education* (1982) 529 F. Supp. 1255, 50 U.S. Law Week 2412)

4. In 1987, in *Edwards v. Aguillard*, the U.S. Supreme Court held unconstitutional Louisiana's "Creationism Act." This statute prohibited the teaching of evolution in public schools, except when it was accompanied by instruction in "creation science." The Court found that, by advancing the religious belief that a supernatural being created humankind, which is embraced by the term *creation science*, the act impermissibly endorses religion. In addition, the Court found that the provision of a comprehensive science education is undermined when it is forbidden to teach evolution except when creation science is also taught. (*1987, 482, U.S. 578, 55 U.S. Law Week 4860, S. CT. 2573, 96 L. Ed. 2d 510*)

5. In 1990, in *Webster v. New Lenox School District*, the Seventh Circuit Court of Appeals found that a school district may prohibit a teacher from teaching creation science, in fulfilling its responsibility to ensure that the First Amendment's establishment clause is not violated, and religious beliefs are not injected into the public school curriculum. The court upheld a district court finding that the school district had not violated Webster's free speech rights when it prohibited him from teaching "creation science," since it is a form of religious advocacy. (*Webster v. New Lenox School District #122*, 917 F. 2d 1004)

6. In 1994, in *Peloza v. Capistrano School District*, the Ninth Circuit Court of Appeals upheld a district court finding that a teacher's First Amendment right to free exercise of religion is not violated by a school district's requirement that evolution be taught in biology classes. Rejecting plaintiff Peloza's definition of a "religion" of "evolutionism", the Court found that the district had simply and appropriately required a science teacher to teach a scientific theory in biology class. (*John E. Peloza v. Capistrano Unified School District*, (1994) 917 F. 2d 1004)

McLean v. Arkansas (1982)

The approach to teaching "creation science" and evolution science" found in Act 590 is identical to the two-model approach espoused by the Institute for Creation Research and is taken almost verbatim from ICR writings. It is an extension of Fundamentalists' view that one must either accept the literal interpretation of Genesis or else believe in the godless system of evolution....

In addition to the fallacious pedagogy of the two model approach, Section 4(2) lacks legitimate educational value because "creation science" as defined in that section is simply not science. Several witnesses suggested definitions of science. A descriptive definition was said to be that science is what is "accepted by the scientific community" and is "what scientists do." The obvious implication of this description is that, in a free society, knowledge does not require the imprimatur of legislation in order to become science.

More precisely, the essential characteristics of science are:

1. It is guided by natural law;

2. It has to be explanatory by reference to natural law;

3. It is testable against the empirical world;

4. Its conclusions are tentative, i.e., are not necessarily the final word; and

5. It is falsifiable. (Ruse and other science witnesses).

Creation science as described in Section 4(a) fails to meet these essential characteristics....

Creation science, as defined in Section 4(a), not only fails to follow the canons defining scientific theory, it also fails to fit the more general descriptions of "what scientists think" and "what scientists do." The scientific community consists of individuals and groups, nationally and internationally, who work independently in such varied fields as biology, paleontology, geology and astronomy. Their work is published and subject to review and testing by their peers. The journals for publication are both numerous and varied. There is, however, not one recognized scientific journal which has published an article espousing the creation science theory described in Section 4(a). Some of the State's witnesses suggested that the

4   VOICES FOR EVOLUTION

scientific community was "close-minded" on the subject of creation-ism and that explained the lack of acceptance of the creation science arguments. Yet no witness produced a scientific article for which publication had been refused. Perhaps some members of the scientific community are resistant to new ideas. It is, however, inconceivable that such a loose knit group of independent thinkers in all the varied fields of science could, or would, so effectively censor new scientific thought.

... The methodology employed by creationists is another factor which is indicative that their work is not science. A scientific theory must be tentative and always subject to revision or abandonment in light of facts that are inconsistent with, or falsify, the theory. A theory that is by its own terms dogmatic, absolutist and never subject to revision is not a scientific theory.

The creationists' methods do not take data, weigh it against the opposing scientific data, and thereafter reach the conclusions stated in Section 4(a). Instead, they take the literal wording of the Book of Genesis and attempt to find scientific support for it....

The Court would never criticize or discredit any person's testimony based on his or her religious beliefs. While anybody is free to approach a scientific inquiry in any fashion they choose, they cannot properly describe the methodology used as scientific, if they start with a conclusion and refuse to change it regardless of the evidence developed during the course of the investigation.

Excerpts from McLean v. Arkansas Board of Education, *529 F. Supp. 1255*

STATE OF TENNESSEE, OFFICE OF THE ATTORNEY GENERAL (1988) Public Schools — Theories of Origins of Life — Creation Science — Establishment Clause

QUESTION:

Whether a teacher in a public school in Tennessee can teach all theories of the origin of life for the purpose of enhancing the effectiveness of science instruction?

OPINION:

It is the opinion of this office that a public school teacher can teach any scientific theory of the origin of life, such as evolution. However, no theory of the origin of life which is religiously based can be taught in the public schools as part of the science curriculum, because its teaching would violate the establishment clause of the First Amendment of the United States Constitution.

ANALYSIS:

The establishment clause of the First Amendment of the United States Constitution provides that "Congress shall make no law respecting an establishment of religion...." Through the Fourteenth Amendment, the Untied States Supreme Court has applied the establishment clause to the states. *See Cantwell v. Connecticut*, 310 U.S. 296 (1940). In determining whether there is a violation of the establishment clause in a particular situation, the Supreme Court, in the case of *Lemon v. Kurtzman*, 403 U.S. 602, 612-613 (1971) announced the following three-prong test:

> First, the legislature must have adopted the law with a secular purpose. Second, the statute's principal or primary effect must be one that neither advances nor inhibits religion. Third, the statute must not result in an excessive entanglement of government with religion.

It should also be noted that the establishment clause applies not only to statutes, but to all actions by public employees and officials which would result in a prohibited promotion of religion. *See Breen v. Runkel*, 614 F. Supp. 355 (W.D. Mich. 1985) (when acting in capacity as classroom instructors, teachers are "state actors" for purpose of determining whether their praying in classrooms, reading from the Bible, and telling stories that have a biblical basis violates the establishment clause.); *Collins v. Chandler Unified School District*, 644 F.2d 759, *cert. denied*, 454 U.S. 863 (1980) (where a high school principal, with the concurrence of their superintendent, granted permission for a student council to recite prayers and Bible verses of their choosing during school hours, there was a violation of the establishment clause).

With regard to your question, a recent decision by the United States Supreme Court held a Louisiana statute that required the teaching of "creation science" in public schools if evolution was taught to be violative of the establishment clause. *Edwards v. Aguillard*, 107 S. Ct. 2573 (1987). In concluding that the statute was unconstitutional, Justice William Brennan, writing for the majority, stated the following with regard to "creation science" as a scientific theory of the origin of life:

> The preeminent purpose of the Louisiana legislature was clearly to advance the religious viewpoint that a supernatural being created human kind. The term 'creation science' was defined as embracing this particular religious doctrine by those responsible for the passage of the Creationism Act. Senator Keith's leading expert on creation science, Edward Boudreaux, testified at the legislative hearings that the theory of creation science included belief in the existence of a supernatural creator.... The legislative history therefore reveals that the term 'creation science' as contemplated by the legislature that adopted this act, embodies the religious belief that a supernatural creator was responsible for the creation of human kind.

Id. at 2581-82. Thus, according to Justice Brennan, "creation science", as understood to include the concept of a supernatural creator, is religiously based and cannot be taught in the public schools as part of the science curriculum without violating the establishment clause.

Justice Brennan's opinion was based upon the record of the legislative debates of the Louisiana statute. No such records exists in this situation. However, the fact that a statute has not been passed in Tennessee requiring the teaching of "creation science" or prohibiting the teaching of evolution unless "creation science" is taught, would not render the actions of a teacher who taught "creation science" as part of the science curriculum to be constitutional. Rather, the teaching of "creation science", if it is intended to include the belief that a supernatural creator was responsible for the creation of life, is an attempt to advance a particular religious view and is violative of the establishment clause of the First Amendment of the United States Constitution.

On the other hand, there would appear to be no constitutional problem with presenting the Biblical account of creation as part of a comparative religion course. *See Abington School District v. Schempp*, 374 U.S. 203, 225 (1963) (Bible may constitutionally be used in an appropriate study of history, civilization, ethics, or comparative religion); *Stone v. Graham*, 449 U.S. 39 (1980) (Ten Commandments cannot be posted on classroom walls but could be discussed in course on ethics).

Opinion no. 88-149
August 18, 1988

WEBSTER V. NEW LENOX
SCHOOL DISTRICT (1990)

I. BACKGROUND*

A. FACTS

Ray Webster teaches social studies at the Oster-Oakview Junior High School in New Lenox, Illinois. In the Spring of 1987, a student in Mr. Webster's social studies class complained that Mr. Webster's teaching methods violated principles of separation between church and state. In addition to the student, both the American Civil Liberties Union and the Americans United for the Separation of Church and State objected to Mr. Webster's teaching practices. Mr. Webster denied the allegations. On July 31, 1987, the New Lenox school board (school board) through its superintendent, advised Mr. Webster by letter that he should restrict his classroom instruction to the curriculum and refrain from advocating a particular religious viewpoint.

Believing the superintendent's letter vague, Mr. Webster asked for further clarification in a letter dated September 4, 1987. In this letter, Mr. Webster also set forth his teaching methods and philosophy. Mr. Webster stated that the discussion of religious issues in his class was only for the purpose of developing an open mind in his students. For example, Mr. Webster explained that he taught nonevolutionary theories of creation to rebut a statement in the social studies textbook indicating that the world is over four billion years old. Therefore, his teaching methods in no way violated the doctrine of separation between church and state. Mr. Webster contended that, at most, he encouraged students to explore alternative viewpoints.

The superintendent responded to Mr. Webster's letter on October 13, 1987. The superintendent reiterated that advocacy of a Christian viewpoint was prohibited, although Mr. Webster could discuss objectively the historical relationship between church and state when such discussions were an appropriate part of the curriculum. Mr. Webster was specifically instructed not to teach creation science, because the teaching of this theory had been held by the federal courts to be religious advocacy.**

Mr. Webster brought suit, principally arguing that the school board's prohibitions constituted censorship in violation of the first and fourteenth amendments. In particular, Mr. Webster argued that the school board should permit him to teach a nonevolutionary theory of creation in his social studies class.

B. THE DISTRICT COURT

The district court concluded that Mr. Webster did not have a first amendment right to teach creation science in a public school. The district court began by noting that, in deciding whether to grant the school district's motion to dismiss, the court was entitled to consider the letters between the superintendent and Mr. Webster because Mr. Webster had attached these letters to his complaint as exhibits. In particular, the district court determined that the October 13, 1987 letter was critical; this letter clearly indicated exactly what conduct the school district sought to proscribe. Specifically, the October 18 letter directed that Mr. Webster was prohibited from teaching creation science and was admonished not to engage in religious advocacy. Furthermore, the superintendent's letter explicitly stated that Mr. Webster could discuss objectively the historical relationship between church and state.

The district court noted that a school board generally has wide latitude in setting the curriculum, provided the school board remains within the boundaries established by the constitution. Because the establishment clause prohibits the enactment of any law "respecting an establishment of religion," the school board could not enact a curriculum that would inject religion into the public schools. U.S. Const. amend. I. Moreover, the district court determined that the school board had the responsibility to ensure that the establishment clause was not violated.

The district court then framed the issue as whether Mr. Webster had the right to teach creation science. Relying on *Edwards v. Aguillard,* 482 U.S. 578 (1987), the district court determined that teaching creation science would constitute religious advocacy in violation of the first amendment and that the school board correctly prohibited Mr. Webster from teaching such material. The court further noted:

Webster has not been prohibited from teaching any nonevolutionary theories or from teaching anything regarding the historical relationship between church and state. Martino's [the superintendent] letter of October 13, 1987 makes it clear that the religious advocacy of Webster's teaching is prohibited and nothing else. Since no other constraints were placed on Webster's teaching, he has no basis for his complaint and it must fail.

Webster v. New Lenox School Dist., Mem. op. at 4-5 (N.D., Ill. May 25, 1989). Accordingly, the district court dismissed the complaint....***

CONCLUSION

For the foregoing reasons, the judgment of the district court is affirmed.

Webster v. New Lenox School District #122, 917 F. 2d 1004
*Introductory material in Background section, preceding the summary of "Facts," is omitted here.
**Footnote in original refers to definition of "creation science" in Edwards v. Aguillard, 482 U.S. 578, 592 (1987)
***Footnote in original omitted here

Peloza v. Capistrano Unified School District (1994)

...C haritably read, Peloza's complaint at most makes this claim: the school district's actions establish a state-supported religion of evolutionism, or more generally of "secular humanism." See complaint at 2-4, 20. According to Peloza's complaint, all persons must adhere to one of two religious belief systems concerning "the origins of life and of the universe": evolutionism, or creationism. Id. at 2. Thus, the school district, in teaching evolutionism, is establishing a state-supported "religion."

We reject this claim because neither the Supreme Court, nor this circuit, has ever held that evolutionism or secular humanism are "religions" for Establishment Clause purposes. Indeed, both the dictionary definition of religion* and the clear weight of the caselaw* are to the contrary. The Supreme Court has held unequivocally that while the belief in a divine creator of the universe is a religious belief, the scientific theory that higher forms of life evolved from lower forms is not. Edwards v. Aguillard, 482 U.S. 578, 96 L. Ed. 2d 510, 107 S. Ct. 2573 (1987) (holding unconstitutional, under Establishment Clause, Louisiana's "Balanced Treatment for Creation-Science and Evolution-Science in Public School Instruction Act").

Peloza would have us accept his definition of "evolution" and "evolutionism" and impose his definition on the school district as its own, a definition that cannot be found in the dictionary, in the Supreme Court cases, or anywhere in the common understanding of the words. Only if we define "evolution" and "evolutionism" as does Peloza as a concept that embraces the belief that the universe came into existence without a Creator might he make out a claim. This we need not do. To say red is green or black is white does not make it so. Nor need we for the purposes of a 12(b)(6) motion accept a made-up definition of "evolution." Nowhere does Peloza point to anything that conceivably suggests that the school district accepts anything other than the common definition of "evolution" and "evolutionism." It simply required him as a biology teacher in the public schools of California to teach "evolution." Peloza nowhere says it required more.

The district court dismissed his claim, stating:

Since the evolutionist theory is not a religion, to require an instructor to teach this theory is not a violation of the Establishment Clause.... Evolution is a scientific theory based on the gathering* and studying of data, and modification of new data. It is an established scientific theory which is used as the basis for many areas of science. As scientific methods advance and become more accurate, the scientific community will revise the accepted theory to a more accurate explanation of life's origins. Plaintiff's assertions that the teaching of evolution would be a violation of the Establishment Clause is [sic] unfounded.

Id. at 12-13. We agree....

John E. Peloza v. Capistrano Unified School District, 37 F. 3d 517
*Footnotes in original are omitted here

PART TWO

Scientific Organizations

ACADEMY OF SCIENCE OF THE ROYAL SOCIETY OF CANADA

The Academy of Science of the Royal Society of Canada considers that "scientific creationism" has nothing to do with science or the scientific method. "Scientific creationism" does not belong in any discussion of scientific principles or theories, and therefore should have no place in a science curriculum.

Science provides knowledge of the natural world in the form of evidence gathered by observation and experiment. Analysis of this evidence allows scientists to generate hypotheses that link and explain different phenomena. Scientific hypotheses must be capable of being tested by further research. If a hypothesis is found to explain many different facts, and even to allow accurate predictions of subsequent discoveries, greater confidence is placed in it, and it is called a theory.

The theory of evolution by natural selection was first clearly formulated in 1859, and for over a century it has been tested and improved by the research of many thousands of scientists: not only by biologists and geologists, but also by chemists and physicists. From deductions based on abundant data, the theory has been developed to explain the changes that have taken place in living things over much of the Earth's history. In its modern form, it remains the only explanation for the diversity of life on this planet that is acceptable to the scientific community.

Science itself evolves, since it must continuously modify existing explanations to incorporate new information. The theory of evolution continues to be refined as new evidence becomes available. Only one thing in science is not open to change: its demand that every explanation be based on observation or experiment, that these be in principle repeatable, and that new evidence be considered.

Scientific creationists adopt an entirely different approach in their attempt to explain the natural world. They accept either biblical or some other authority as overriding other kinds of evidence. They reject much of the accumulated scientific knowledge, and commonly deny the validity of deductions based on directly observable phenomena such as radioactive decay. This is because their philosophy is rooted in a different aspect of human culture. If their claim, that the Earth and all its living things were created only several thousand years ago, was correct, many of the central concepts of modern science would have to be abandoned. The methodology and conclusions of scientists and "scientific creationists" are therefore incompatible, and the term "scientific creationism" is a contradiction in terms, since it has no basis in science.

Delivered by Fellows of the Academy to each Provincial Minister of Education in Canada. Published in Geotimes, *November 1985, p. 21.*

ALABAMA ACADEMY OF SCIENCE (1981)

The Executive Committee of the Alabama Academy of Science hereby records its opposition to legislation to introduce "scientific creationism" into the Alabama classroom. Furthermore, the Executive Committee of the Alabama Academy of Science believes that the introduction of classroom subject content through the political process not only violates the academic freedom of the subject specialist to determine relevant and scientifically sound concepts, but also represents an inappropriate and potentially dangerous precedent for American education.

Adopted by a vote of 24 in favor to 7 opposed; copy hand-dated 1981.

ALABAMA ACADEMY
OF SCIENCE, INC. (1994)

The Executive Committee of the Alabama Academy of Science strongly deplores efforts to insert into the Course of Study for Science for the public schools of this state theories and hypotheses which do not meet the cardinal criteria of the hypotheses, theories and laws of science: that they be based on facts and that they be capable of being proven false. To be scientific, hypotheses, theories and laws must be in accord with the results of repeatable controlled experiments or be formulated as the result of consistent and verifiable observations.

Adopted by the Executive Committee October 29, 1994

AMERICAN
ANTHROPOLOGICAL ASSOCIATION

W*hereas* evolutionary theory is the indispensable foundation for the understanding of physical anthropology and biology;

Whereas evolution is a basic component of many aspects of archeology, cultural anthropology, and linguistics;

Whereas evolution is a basic component of allied disciplines such as the earth sciences and a cornerstone of 20th-century science in general;

Whereas a century of scientific research has confirmed the reality of evolution as a historical process, and the concept of evolution, in all its diversity, has explained the scientifically known evidence and successfully predicted fruitful paths of further research; and

Whereas local and national campaigns by so-called scientific Creationists and other antievolutionists nevertheless challenge the right of public schools to teach evolutionary theory without giving scientific credence or equal time to Creationist and other antievolutionist explanations of the origin and development of life;

Be it moved that the American Anthropological Association affirms the necessity of teaching evolution as the best scientific explanation of human and nonhuman biology and the key to understanding the origin and development of life, because the principles of evolution have been tested repeatedly and found to be valid according to scientific criteria;

The Association respects the right of people to hold diverse religious beliefs, including those which reject evolution, as matters of theology or faith but not as tenets of secular science;

Efforts to require teaching Creationism in science classes, whether exclusively, as a component of science curricula, or in equal-time counterpoint to evolution, are not based on science but rather are attempts to promote unscientific viewpoints in the name of science without basis in the record of scientific research by generations of anthropologists and other scholars;

The subject of life origins is addressed in tremendous diversity among the world's religions, and efforts to promote particular Judeo-Christian creation accounts in public schools are ethnocentric as well as unscientific.

Be it further moved that the Association shall communicate this motion upon passage to the public news media, to commissioners of education or equivalent officials in each of the 50 states, and to other officials and organizations deemed appropriate by the Executive Board or Executive Director.

Be it further moved that members of the Association are encouraged to promote these points of professional concern in their home communities among educators, parents, and students and in appropriate public forums beyond the boundaries of traditional, professional, and academic disciplines.

Passed at 1980 annual meeting in Washington, DC.

AMERICAN ASSOCIATION FOR THE ADVANCEMENT OF SCIENCE
A Statement on the Present Scientific Status of the Theory of Evolution (1923)

*I*nasmuch as the attempt has been made in several states to prohibit in tax-supported institutions the teaching of evolution as applied to man, and

Since it has been asserted that there is not a fact in the universe in support of this theory, that it is a "mere guess" which leading scientists are now abandoning, and that even the American Association for the Advancement of Science at its last meeting in Toronto, Canada, approved this revolt against evolution, and

Inasmuch as such statements have been given wide publicity through the press and are misleading public opinion on this subject,

Therefore, the council of the American Association for the Advancement of Science has thought it advisable to take formal action upon this matter, in order that there may be no ground for misunderstanding of the attitude of the association, which is one of the largest scientific bodies in the world, with a membership of more than 11,000 persons, including the American authorities in all branches of science. The following statements represent the position of the council with regard to the theory of evolution.

1. The council of the association affirms that, so far as the scientific evidences of the evolution of plants and animals and man are concerned, there is no ground whatever for the assertion that these evidences constitute a "mere guess." No scientific generalization is more strongly supported by thoroughly tested evidences than is that of organic evolution.

2. The council of the association affirms that the evidences in favor of the evolution of man are sufficient to convince every scientist of note in the world, and that these evidences are increasing in number and importance very year.

3. The council of the association also affirms that the theory of evolution is one of the most potent of the great influences for good

that have thus far entered into human experience; it has promoted the progress of knowledge, it has fostered unprejudiced inquiry, and it has served as an invaluable aid in humanity's search for truth in many fields.

4. The council of the association is convinced that any legislation attempting to limit the teaching of any scientific doctrine so well established and so widely accepted by specialists as is the doctrine of evolution would be a profound mistake, which could not fail to injure and retard the advancement of knowledge and of human welfare by denying the freedom of teaching and inquiry which is essential to all progress.

Resolution adopted 1923

AMERICAN ASSOCIATION FOR THE ADVANCEMENT OF SCIENCE (1972)

W *hereas* the new Science Framework for California Public Schools prepared by the California State Advisory Committee on Science Education has been revised by the California State Board of Education to include the theory of creation as an alternative to evolutionary theory in discussions of the origins of life, and

Whereas the theory of creation is neither scientifically grounded nor capable of performing the roles required of scientific theories, and

Whereas the requirement that it be included in textbooks as an alternative to evolutionary theory represents a constraint upon the freedom of the science teacher in the classroom, and

Whereas its inclusion also represents dictation by a lay body of what shall be considered within the corpus of a science,

Therefore we, the members of the Board of Directors of the American Association for the Advancement of Science, present at the quarterly meeting of October 1972, strongly urge that the California State Board of Education not include reference to the theory of creation in the new Science Framework for California Public Schools and that it adopt the original version prepared by the California State Advisory Committee on Science Education.

22 October 1972.

American Association for the Advancement of Science (1982)
Forced Teaching of Creationist Beliefs in Public School Science Education

W hereas it is the responsibility of the American Association for the Advancement of Science to preserve the integrity of science, and

Whereas science is a systematic method of investigation based on continuous experimentation, observation, and measurement leading to evolving explanations of natural phenomena, explanations which are continuously open to further testing, and

Whereas evolution fully satisfies these criteria, irrespective of remaining debates concerning its detailed mechanisms, and

Whereas the Association respects the right of people to hold diverse beliefs about creation that do not come within the definitions of science, and

Whereas Creationist groups are imposing beliefs disguised as science upon teachers and students to the detriment and distortion of public education in the United States

Therefore be it resolved that because "Creationist Science" has no scientific validity it should not be taught as science, and further, that the AAAS views legislation requiring "Creationist Science" to be taught in public schools as a real and present threat to the integrity of education and the teaching of science, and

Be it further resolved that the AAAS urges citizens, educational authorities, and legislators to oppose the compulsory inclusion in science education curricula of beliefs that are not amenable to the process of scrutiny, testing, and revision that is indispensable to science.

The above resolution is a composite of draft resolutions written by D. Allen Bromley, Edward R. Brunner, Anna J. Harrison, and Glynn Isaac. It was passed by the AAAS Board of Directors on 4 January 1982 and submitted to the Council as a proposed joint resolution of the Board and Council. It was passed by Council on 7 January, and published in Science 215:1072 on 26 February.

American Association for the Advancement of Science (Commission on Science Education)

The Commission on Science Education of the American Association for the Advancement of Science is vigorously opposed to attempts by some boards of education and other groups to require that religious accounts of creation be taught in science classes.

During the past century and a half, the earth's crust and the fossils preserved in it have been intensively studied by geologists and paleontologists. Biologists have intensively studied the origin, structure, physiology, and genetics of living organisms. The conclusion of these studies is that the living species of animals and plants have evolved from different species that lived in the past. The scientists involved in these studies have built up the body of knowledge known as the biological theory of the origin and evolution of life. There is no currently acceptable alternative scientific theory to explain the phenomena.

The various accounts of creation that are part of the religious heritage of many people are not scientific statements or theories. They are statements that one may choose to believe, but if he does, this is a matter of faith, because such statements are not subject to study or verification by the procedures of science. A scientific statement must be capable of test by observation and experiment. It is acceptable only if, after repeated testing, it is found to account satisfactorily for the phenomena to which it is applied.

Thus the statements about creation that are part of many religions have no place in the domain of science and should not be regarded as reasonable alternatives to scientific explanations for the origin and evolution of life.

Adopted by the Commission on Science Education of the AAAS at its meeting on 13 October 1972 in Washington, DC.

AMERICAN ASSOCIATION OF PHYSICAL ANTHROPOLOGISTS

1. *Be it resolved* that the American Association of Physical Anthropologists strongly endorses the recent resolution of the American Association for the Advancement of Science condemning the concepts of and teaching of, at public expense, so-called scientific creationism. *

2. *Whereas* the American Association of Physical Anthropologists recognizes the advantages to any society which accrue when its members accept some moral code of behavior, and

   Whereas the Association supports the Constitutional provision separating church and state,

   Therefore be it resolved that the Association condemns any effort by the state to dictate specific religious instruction to the people, and

   Be it further resolved that the Association condemns any effort by the state or any group within the state to restrict the right of all individuals to freedom of religious expression by advancing one religious viewpoint.

3. *Whereas* the American Association of Physical Anthropologists recognizes that our modern society is based on a high degree of technological and scientific sophistication, and

   Whereas the Association realizes that such technology and science can only be sustained if there is continuous advancement in our knowledge of and control over natural phenomena, and

   Whereas such continuous advancement can only be sustained if instruction in the current state of knowledge be available to all our citizens, and

   Whereas public understanding of our technological society, which will promote the individual's ability to cope and serve, can only be achieved if instruction in the sciences reflects the current content of scientific research,

   Be it resolved that the American Association of Physical Anthropologists charges the state with the duty of providing, through the public education system, the people with instruction in the current state of objective knowledge concerning our natural universe.

4. Be it resolved that the Secretary is directed to communicate these three resolutions to as many individuals or organizations as possible who may be concerned with these issues.

1982
See the 1982 statement by the American Association for the Advancement of Science on page 25

AMERICAN ASTRONOMOCIAL SOCIETY
Resolution on Creationism

During the past year, religious fundamentalists have intensified their effort to force public school science classes to include instruction in "creationism." As defined in publications of the Institute for Creation Research and in laws passed or under consideration by several state legislatures, this doctrine includes the statement that the entire universe was created relatively recently, i.e. less than 10,000 years ago. This statement contradicts results of astronomical research during the past two centuries indicating that some stars now visible to us were in existence millions or billions of years ago, as well as the results of radiometric dating indicating that the age of the earth is about 4.5 billion years.

The American Astronomical Society does not regard any scientific theory as capable of rigorous proof or immune to possible revision in the light of new evidence. Such evidence should be presented for critical review and confirmation in the appropriate scientific journals. In this case, no such evidence for recent creation of the earth and universe has survived critical scrutiny by the scientific community. It would therefore be most inappropriate to demand that any science teacher present it as a credible hypothesis.

We agree with the findings of Judge William Overton that the Arkansas creationism law represents an unconstitutional intrusion of religious doctrine into the public schools, that "creation science" is not science, and that its advocates have followed the unscientific procedure of starting from a dogmatically held conclusion and looking only for evidence to support that conclusion.

The American Astronomical Society deplores the attempt to force creationism into public schools and urges Congress, all state legislatures, local school boards and textbook publishers to resist such attempts.

Adopted unanimously on 10 January 1982

AMERICAN CHEMICAL SOCIETY

ddendum to Report of Committee on Professional and Member Relations:

There is increased pressure on boards of education to mandate the teaching of biblical creationism in the nation's public school science classes. As recent examples of this pressure, the state legislatures of Arkansas and Louisiana have passed measures requiring that such creationism be taught whenever biological (Darwinian) evolution is taught.

The Board of Directors of the American Chemical Society reaffirms its statement of December 2, 1972 that creationism theories, often mistermed "scientific creationism," should not be taught as science in the nation's science classes. These theories were not derived from scientific data and are not amenable to scientific test. Any implication that such theories are within the framework of science would confuse students about the nature of both religion and science.

AMERICAN GEOLOGICAL INSTITUTE

S cientific evidence indicates beyond any doubt that life has existed on Earth for billions of years. This life has evolved through time producing vast numbers of species of plants and animals, most of which are extinct. Although scientists debate the mechanism that produced this change, the evidence for the change is undeniable. Therefore, in the teaching of science we oppose any position that ignores this scientific reality, or that gives equal time to interpretations based on religious beliefs only.

Unanimously approved by the governing board on 5 November 1981.

AMERICAN GEOPHYSICAL UNION

The Council of the American Geophysical Union notes with concern the continuing efforts by creationists for administrative, legislative, and juridical actions designed to require the teaching of creationism as a scientific theory.

The American Geophysical Union is opposed to all efforts to require the teaching of creationism or any other religious tenets as science.

Passed unanimously by the AGU Council on 6 December 1981

AMERICAN INSTITUTE
OF BIOLOGICAL SCIENCES

The AIBS Executive Committee passed a resolution in 1972 deploring efforts by Biblical literalists to interject creationism and religion into science courses. It is very troubling that more than 20 years later, there is an urgent need to reaffirm AIBS's earlier position. Despite rulings by the Supreme Court declaring it unconstitutional to promote a religious perspective in public school education, such attempts by creationists continue in a variety of guises.

The theory of evolution is the only scientifically defensible explanation for the origin of life and development of species. A theory in science, such as the atomic theory in chemistry and the Newtonian and relativity theories in physics, is not a speculative hypothesis, but a coherent body of explanatory statements supported by evidence. The theory of evolution has this status. The body of knowledge that supports the theory of evolution is ever growing: fossils continue to be discovered that fill gaps in the evolutionary tree and recent DNA sequence data provide evidence that all living organisms are related to each other and to extinct species. These data, consistent with evolution, imply a common chemical and biological heritage for all living organisms and allow scientists to map branch points in the evolutionary tree.

Biologists may disagree about the details of the history and mechanisms of evolution. Such debate is a normal, healthy, and necessary part of scientific discourse and in no way negates the theory of evolution. As a community, biologists agree that evolution occurred and that the forces driving the evolutionary process are still active today. This consensus is based on more than a century of scientific data gathering and analysis.

Because creationism is based almost solely on religious dogma stemming from faith rather than demonstrable facts, it does not lend itself to the scientific process. As a result, creationism should not be taught in any science classroom.

Therefore, AIBS reaffirms its 1972 resolution that explanations for the origin of life and the development of species that are not supportable on scientific grounds should not be taught as science.

Board Resolution 1994

American Physical Society

The Council of the American Physical Society opposes proposals to require "equal time" for presentation in public school science classes of the biblical story of creation and the scientific theory of evolution. The issues raised by such proposals, while mainly focused on evolution, have important implications for the entire spectrum of scientific inquiry, including geology, physics, and astronomy.

In contrast to "Creationism," the systematic application of scientific principles has led to a current picture of life, of the nature of our planet, and of the universe which, while incomplete, is constantly being tested and refined by observation and analysis. This ability to construct critical experiments, whose results can require rejection of a theory, is fundamental to the scientific method.

While our society must constantly guard against oversimplified or dogmatic descriptions of science in the education process, we must also resist attempts to interfere with the presentation of properly developed scientific principles in establishing guidelines for classroom instruction or in the development of scientific textbooks.

We therefore strongly oppose any requirement for parallel treatment of scientific and non-scientific discussions in science classes. Scientific inquiry and religious beliefs are two distinct elements of the human experience. Attempts to present them in the same context can only lead to misunderstandings of both.

Published as a news release dated 15 December 1981 on letterhead of the American Institute of Physics. The APS describes itself in this release as "the largest professional society of physicists in America, with more than 32,000 members."

American Psychological Association

Principles of evolution are an essential part of the knowledge base of psychology. Any attempt to limit or exclude the teaching of evolution from the science curriculum would deprive psychology students of a significant part of their education.

Currently, groups identifying themselves as "creationists" are proposing legislation to require teaching of "creation science" as part of the science curriculum of public schools.

The American Psychological Association, without questioning the right of any individual to hold "creationist" beliefs, views "creationism" as a set of religious doctrines that do not conform to criteria of science. Scientific views are empirically testable, continually open to the processes of scrutiny and experimentation that are the essence of science.

The American Psychological Association believes that "creationism" does not meet the criteria of science and should not be taught as part of the public school science curriculum. Further, the American Psychological Association is opposed to any attempts to require by statute or other means the inclusion of "creationism" within the science curriculum of the public schools.

Passed by a vote of 100 in favor to 1 opposed at the APA annual meeting, 1982.

AMERICAN SOCIETY
OF BIOLOGICAL CHEMISTS

Evolutionary theory is concerned with certain past, present, and future biological events. Like other scientific hypotheses, it leads to predictions, many but not all of which are subject to experimental observation and scientific tests. Evolutionary theory is compatible with many, but not all, religious beliefs; by itself it is not, was not meant to be, and should never be presented as a religious belief. Its proper forum is the science classroom.

The term "Creation-Science" obscures the profound differences between religious beliefs and scientific theory. The proper education of the nation's youth for citizenship in a technological age demands that the distinction between these two major currents in human affairs be maintained in keeping with the precepts of our Constitution.

---

25 August 1982. Ballot referendum approved by the ASBC membership by vote of 2624 in favor to 151 opposed. Reported to membership in a memorandum of 30 November 1981 by Charles C. Hancock, Executive Officer.

AMERICAN SOCIETY OF PARASITOLOGISTS:
Society Takes Stand on Creationism

The American Society of Parasitologists — a national membership organization of 1500 professional scientists — vigorously opposes any state or federal law or any public school board policy that would diminish public education on the principle of evolution, or that would demand comparable funding or treatment of creationism. Some of the society's grounds for this opposition are:

1. CREATIONISM IS NOT A SCIENCE AND CANNOT BECOME A SCIENCE

Science is a disciplined method of obtaining naturalistic explanations of the world and universe. God is believed to exist outside the domain of natural law and to transcend its limitations. Creationism inherently rests on belief in this supernatural Creator, and no supernatural premise can ever be correctly considered a science.

2. EVOLUTION IS NOT ANTI-CHRISTIAN OR ANTI-RELIGIOUS

Science makes no pretense of judging whether or not God exists or why He works as He does; science has always acknowledged these questions as being outside the domain of its authority. In their private beliefs, many, perhaps the majority, of scientists who believe the principle of evolution are also God-believing Christians, Jews, Moslems, or other theists, and see no contradiction between these beliefs. Many, for example, see evolution as God's mechanism of ongoing creation. Furthermore, the official positions enunciated by American and world leaders of Roman Catholic, Episcopal, Presbyterian, and other churches are that evolution is not a contradiction of Biblical religion. They opine that the Judeo-Christian creation story is "a religious myth system ... neither empirical science nor recorded history, [but] a religious interpretation divinely inspired in a prescientific age."

3. FUNDAMENTALIST RELIGION IS THE SOLE REASON FOR THE CREATIONIST CAUSE

When the U.S. Supreme Court struck down Arkansas' creationist law in 1968, Justice Fortas ruled that the Arkansas law could not be justified on the grounds of any state policy "other than the religious views of some of its citizens. It is clear that fundamentalist sectarian conviction was and is the law's reason for existence." This is equally true today and the appellation "scientific creationism" cannot disguise that basic intent (see also the ruling of U.S. District Court Judge William R. Overton, in the recent Arkansas trial on creationism in schools published in *Science* 215:934-943, 1982). Neither science nor public education has any interest in or potential benefit from the passage of such laws, which exist only to benefit a certain denomination of Christians. The 123-year history of creationism clearly shows it to be tied to no other cause but this, and to be overwhelmingly rejected by the majority of Christian denominations and by scientists of all faiths.

4. CREATIONISM INFRINGES ON THE UNITES STATES CONSTITUTION

Because creationism is linked solely with fundamentalist Christianity, all creationist laws infringe on the First Amendment clause prohibiting the establishment of religion. Current creationist bills also infringe on the due process clause of the Fourteenth Amendment which has been judged to imply that no law is constitutional which is too vague or ambiguous to be reasonably obeyable. Creationist bills require instruction in creationism yet prohibit instruction in any religious doctrine. Creationism necessarily implies a supernatural creator, and this is necessarily a religious concept. Creationist laws are therefore unconstitutionally ambiguous or self-contradictory. Instruction in evolution is not unconstitutional despite the claims of creationists that it is so. Evolution has a scientific not a religious basis and is believed by nearly all professional life scientists regardless of their religious beliefs. Evolution does not violate the free exercise clause of the First Amendment, for scientific education in evolution does not prohibit the student from being taught otherwise in the home and church.

5. THE BUSINESS OF THE SCIENCE CURRICULUM IS ONLY TO TEACH PREVAILING SCIENTIFIC VIEWPOINTS

Any public school science course must cover a large body of knowledge in a short academic term, and is necessarily limited to teaching only those views which are well established and widely accepted by the scientific community. The fact that some scientists reject evolution does not warrant inclusion of their views in lower-level science curricula. There are many minority beliefs in science besides creationism that are excluded from consideration or from presentation as valid scientific fact or theory. The scientific community is inherently and traditionally vigorous in its criticism of established beliefs and introduction of new concepts. If the anti-Darwinian views of fundamentalists have any validity as science, they will eventually become widely accepted. If so it will be on their scientific and not their religious merit. Only then will they warrant treatment in the public school curriculum.

6. CREATIONISM IS AN INFRINGEMENT OF ACADEMIC FREEDOM

Science teachers are already free to mention or discuss creationism in the classroom if they wish, so long as they do not materially compromise the educational objective of the schools to cover the major areas of scientific information. To legislate creationism infringes on the rights of those teachers, students, and parents who believe the curriculum must be religiously neutral and that non-science does not belong in the science class.

7. EVOLUTION IS FACTUAL AND ESSENTIAL TO BIOLOGICAL EDUCATION

The word "theory" has different meanings to the scientist and layman. Virtually all scientists accept the evolution of current species from fewer, simpler, ancestral ones as undisputed fact. The "theory" of evolution pertains merely to the mechanisms by which this occurs, and the much-touted arguments among scientists about evolution are over details of these mechanisms, not about the factuality of evolution itself. To call evolution a theory implies no more doubt about its factuality than referring to atomic theory or the theory of gravitation means we doubt the existence of atoms or gravity. To excise evolution from the biology curriculum would reduce biology courses to a series of disconnected facts and severely inhibit those aspects of the discipline which contribute to creative scholarship.

The above statement is a composite of drafts by Walter M. Kemp and Kenneth S. Saladin, adopted by the ASP Council and published in the ASP Newsletter 4(1):6-8 in March 1982.

CALIFORNIA ACADEMY OF SCIENCES
A Statement Affirming the Central Role of Scientific Principles in the Teaching of Evolutionary Biology

Evolutionary biology, like every other natural science, is a powerful expression of human curiosity and intellect. With techniques for reconstructing the history of life on Earth, Homo sapiens has become uniquely capable of knowing about its own past as well as that of other organisms on this planet. Discoveries in phylogenetics, paleontology, genetics, and developmental and molecular biology give us the capacity to test our theories and to develop new ones, using a vast store of empirical data and increasingly sophisticated methods. Continued opportunity to perform such tests has resulted in further support for descent with modification, justifying the fundamental role that evolution plays in our understanding of humanity's place in nature. It provides a rational basis for dealing with such problems as preserving the quality of our environment, and enhancing the quality of our lives.

Now, more than ever, is a time when intellectual standards need to be upheld. For example, it is crucial that we clearly distinguish between such legitimate natural sciences as astronomy and such pseudosciences as astrology. There is a fundamental difference between testing hypotheses so as to reject some in favor of alternatives, and rationalization in terms of a dogmatic belief system.

The natural sciences have a long history of weeding out notions inherited from pre-scientific culture, often in the face of determined resistance. Repeatedly, old arguments, long since refuted, have been refurbished and presented to new audiences that are ill-equipped to evaluate them. Lately, creationist pseudoscience has been attempting to insinuate itself into the curriculum under the rubric of "intelligent design." Prior to the fundamental contribution of Darwin in 1859, there seemed to be no way to explain the remarkable adaptations of organisms except in terms of a miracle. With the discovery and recognition of natural selection, this argument was shown to depend upon a pre-Darwinian failure of the human imagination to find testable, scientific explanations for the origin and diversity of life. The appropriate place in the science curriculum for the notion that organisms have been designed is the same as that for the notion that the earth is located at the center of the universe.

Science and religion are concerned with different aspects of human life and are evaluated according to fundamentally different criteria. Failing to make this distinction gives the false impression that we are limited to two alternatives when faced with an apparent contradiction.

Insofar as belief in special creation is a part of many religions, it needs to be understood in the context of the comparative and historical study of culture. Religion has played and continues to play an important role in human life, and our citizens need to be well informed about it. In recognizing the rich cultural diversity of beliefs and practices both past and present, schools should teach about all religions, provided that this is done in a fair and objective manner, without proselytizing. All this can be accomplished without compromising the central role that scientific principles must take in the teaching of evolutionary biology.

Adopted unanimously by Curator's Forum, November 16, 1994
Passed by Science Council, November 30, 1994

GEOLOGICAL SOCIETY OF AMERICA

The Geological Society of America believes in the importance of using scientific documentation and reasoning. Biological evolution is a particularly impressive example of a principle derived in this way; we geologists find incontrovertible evidence in the rocks that life has existed here on Earth for several billions of years and that it has evolved through time. Although scientists debate the mechanism that produced this change, the evidence for the change itself is undeniable.

The ideas of "creationism," on the other hand, lack any similar body of supporting evidence. We oppose including creationism in science courses in public schools on the grounds that its conclusions were not obtained using scientific methods. Creationism weakens the emphasis on scientific reasoning that is essential to the continued advancement of scientific knowledge.

Drafted by GSA Councilors Rosemary J. Vidale, Maria Luisa B. Crawford, and Peter J. Wyllie, and adopted by the Council at its May 1983 meeting. Published in GSA News and Information, November 1983, p. 177.

GEORGIA ACADEMY OF SCIENCE (1980)

W*hereas* members of the Georgia Academy of Science are duly trained in their respective scientific disciplines by years of education and experience, and

Whereas members of the Georgia Academy of Science have considered creationism in light of their scientific experience and religious beliefs, and

Whereas members of the Georgia Academy of Science have the following concerns about creationism:

1. Philosophically, "scientific creationism" or "divine creationism" is not based upon objectively-gathered data and testing of the model as required by science.

2. Legally, the required teaching of "creationism" might violate the separation of religion and state. It would definitely establish precedent for the legal inclusion of creation narratives of many religions into the science curriculum. The precedent would also be set for other groups to make demands for modifications in the curriculum of disciplines other than science.

3. Pedagogically, problems could result by requiring science teachers to teach as science a model of divine creationism in which they have not been trained. Moreover, various local groups might demand that divine creation be taught according to their own religious beliefs.

Be it, therefore, resolved that the members of the Georgia Academy of Science oppose the teaching of "creationism" in the science curriculum.

Passed unanimously by plenary session of the Georgia Academy of Science on 19 April 1980.

GEORGIA ACADEMY OF SCIENCE (1982)

SYNOPTIC POSITION STATEMENT OF THE GEORGIA ACADEMY
OF SCIENCE WITH RESPECT TO THE FORCED TEACHING OF
CREATION-SCIENCE IN PUBLIC SCHOOL SCIENCE EDUCATION

The great majority of scientists and teachers of science in the primary schools, high schools, colleges, and universities of Georgia are both evolutionists and Christians, or Jews, or adherents to some other religious preference. A few may adhere to no religion. In a pluralistic society students represent a comparable religious spectrum.

Based upon overwhelming scientifically verifiable evidence to date, most scientists, regardless of religious preference, think that the earth and all forms of life evolved over a period of several billion years. Evolution can be viewed as a creative process continuing over long periods of time. The extensive evidence of evolution is not in opposition to the variety of religious concepts or creation by a supreme being. The causative beginning of primeval appearance of matter or life in our universe is not at issue. The evidence of evolution does not claim to reveal the primal source of energy, matter, or life. The latter is a question which is addressed by the various religions outside the walls of our publicly funded educational institutions.

On January 5, 1982, U.S. Circuit Court Judge William R. Overton ruled Arkansas' "Balanced Treatment for Creation-Science and Evolution-Science" Act to be a violation of the constitutional separation of church and state. The Act had the advancement of religion as its primary goal, in his opinion. A month later, the attorney general of Arkansas announced his decision not to appeal Overton's opinion because the state had little chance of winning in higher federal court. The plaintiffs in this landmark case included components of the Southern Baptist, Presbyterian, United Methodist, Episcopal, and Roman Catholic churches, in addition to the American Jewish Congress, and the Union of Hebrew Congregations. Other plaintiffs included the Arkansas Education Association, the National

Association of Biology Teachers, and the National Coalition for Public Education and Religious Liberty.

The Georgia Academy of Science concurs with the following resolution adopted in January of 1982 by the American Association for the Advancement of Science (AAAS) pertaining to the Forced Teaching of Creationist Beliefs in Public School Science Education: *[See the 1982 statement by the American Association for the Advancement of Science on page 25.]*

IOWA ACADEMY OF SCIENCE (1982)

C urrent attempts to introduce "scientific creationism" into the science classroom are strongly opposed by The Iowa Academy of Science on the grounds that creationism when called "scientific" is a religious doctrine posed as science. It is contrary to the nature of science to propose supernatural explanations of natural events or their origins. With its appeal to the supernatural, creationism is outside the realm of science.

Creationist organizations that are advocating the teaching of "scientific creationism" in science classrooms include members purported to be scientists who have examined the evidence and have found creationism to be a superior alternative to evolution. They claim to know of evidence that supports the idea of a young earth and that shows evolution to be impossible. Much of this "evidence" is inaccurate, out of date, and not accepted by recognized paleontologists and biologists. The total membership of these "scientific" creationist groups constitutes only a fraction of one percent of the scientific personnel in this country. Most of them are not trained in biology or geology, the areas in which professional judgments are made in the field of evolutionary theory. They often misrepresent the positions of respected scientists and quote them out of context to support their own views before audiences and government bodies. They are driven by the notion that all explanations of natural events must conform to their preconceived creationist views. These tactics are used to give the uninformed public the false impression that science itself is confused. Then a supernatural explanation is proposed to bring order out of apparent chaos.

The Iowa Academy of Science urges legislators, school administrators, and the general public not to be misled by the tactics of these so-called "scientific creationists." The Academy respects the right of persons to hold diverse religious beliefs, including those which reject evolution, but only as matters of theology or faith, not as secular science. Creationism is not science and the Academy deplores and opposes any attempt to disguise it as science. Most recognized scientists find no conflict between religious faith and acceptance of evolution. They do not view evolution as being anti-religious. They have no vested interest in supporting evolution as do the "scientific creationists" in supporting creationism, but merely consider evolution as being most consistent with the best evidence.

The Iowa Academy of Science feels strongly that the distinction between science and religion must be maintained. A state with one of the highest literacy rates and with the highest scientific literacy scores in the nation, and one which prides itself on the individuality of its citizens, should discriminate in its public education system between what is science and what is not science.

Approved by a majority of all voting members of the Iowa Academy of Science in February, 1981.

STATEMENT OF THE POSITION OF THE IOWA ACADEMY OF SCIENCE ON PSEUDOSCIENCE (1986)

The Iowa Academy of Science strongly opposes the public promotion of pseudoscience, whether through the media, the legislature, or classrooms of accredited educational institutions of Iowa.

"Pseudoscience" is a catch-all term for any mistaken or unsupported beliefs that are cloaked in the disguise of scientific credibility. Examples include assertions of scientific creationism, the control of actions at a distance through meditation, and the belief in levitation, astrology, or UFO visitors. While the IAS opposes the promotion of such beliefs, it does not oppose critical examination of them, either in the public media or in classrooms. Indeed, there is much to be learned from critical examination of pseudoscience.

One main concern is public confusion over what science is and what it is not. This cannot be resolved merely by contriving tighter definitions of science or its methods. In fact, authoritative definitions inadvertently provide a model that counterfeiters need in order to better fashion their "cloaks of scientific credibility". To clear up the confusion between real and bogus science we must focus not on their definitions, but on their differences.

In contrast to pseudoscientists, scientists seek out, expose, and correct any logical fallacies or other errors which could weaken their theories or interpretations. To assure complete scrutiny, open criticism is not only tolerated but often rewarded, particularly when it results in significant revisions of established views. The debate is held in refereed scientific journals and in meetings, and anyone, well-known or not, can submit pro or con arguments for publication or presentation before peers.

By contrast, open criticism is not welcomed by pseudoscientists. They usually avoid publishing in refereed scientific journals, and subsequently their theories are not self-correcting; thus they fail to experience the progressive changes characteristic of science. Astrology and creationism, for example, have experienced nothing comparable to Copernican or Darwinian revolutions (paradigm shifts) which have occurred in astronomy and biology.

The Iowa Academy of Science is prepared to assist citizens, teachers, public officials and the media who seek information on issues involving science and pseudoscience.

KENTUCKY ACADEMY OF SCIENCE

The Kentucky Academy of Science is opposed to any attempt by legislative bodies to mandate the specific content of science courses. The content of science courses should be determined by the standards of the scientific community. Science involves a continuing systematic inquiry into the manifold aspects of the biological and material world. It is based upon testable theories which may change with new data; it cannot include interpretations based on faith or religious dogma. As scientists we object to attempts to equate "scientific creationism" and evolution as scientific explanations of events. Teaching the so-called "two model" approach would not only imply that these views are equivalent alternatives among scientists, it would also be misleading to students. The two "models" are not equivalent. There is overwhelming acceptance by scientists of all disciplines that evolution (the descent of modern species of animals and plants from different ancestors that lived millions of years ago) is consistent with the weight of a vast amount of evidence. The understanding of the processes underlying evolution has provided the foundation upon which many of the tremendous advances in agriculture and medicine and theoretical biology have been built. Differences among scientists over questions of how evolution was accomplished do not obscure the basic agreement that evolution has occurred.

Most people who subscribe to religious views have developed belief systems that are compatible with evolution. There is a widespread consensus among theologians that biblical accounts of creation are misunderstood if they are treated as literal scientific explanations. We fully respect the religious views of all persons but we object to attempts to require any religious teachings as science.

We join the National Academy of Sciences, the American Association for the Advancement of Science and the academies of science in many other states in calling for the rejection of attempts to require the teaching of "scientific creationism" as a scientific theory.

It is further recommended that the Kentucky Academy of Science encourage its members and other professional scientific groups to give support and aid to those classroom teachers who present the subject matter of evolution fairly and encounter community objection. We also encourage administrators and individual teachers to oppose the inclusion of nonscientific concepts in the science classroom.

1981, revised 1983

Louisiana Academy of Sciences

Whereas the stated goal of the Louisiana Academy of Sciences is to encourage research in the sciences and disseminate scientific knowledge, and

Whereas such pursuits are based on the scientific method requiring the testing of hypotheses before their inclusion in the body of scientific knowledge, and

Whereas organic evolution is amenable to repeated observation and testing, and

Whereas the ideas of creation are not amenable to verification by observation and experimentation, and

Whereas the Academy respects and supports the right of people to possess beliefs in creation and other matters that are not encompassed by the subject matter of science,

Therefore be it resolved that the terms "creation science" or "scientific creationism" are artificial and have been used to refer to purported areas of knowledge that do not exist, and

Be it also resolved that the members of the Louisiana Academy of Sciences urge fellow Louisianans, political leaders, and educators to oppose the inclusion in state science programs of the so-called discipline of creation science or other similar ideas which cannot be tested, accepted, or rejected by the scientific method.

Passed by the general membership at the annual meeting on 5-6 February 1982.

NATIONAL ACADEMY OF SCIENCES (1972)

W*hereas* we understand that the California State Board of Education is considering a requirement that textbooks for use in the public schools give parallel treatment to the theory of evolution and to belief in special creation; and

Whereas the essential procedural foundations of science exclude appeal to supernatural causes as a concept not susceptible to validation by objective criteria; and

Whereas religion and science are, therefore, separate and mutually exclusive realms of human thought whose presentation in the same context leads to misunderstanding of both scientific theory and religious belief; and

Whereas, further, the proposed action would almost certainly impair the proper segregation of teaching and understanding of science and religion nationwide, therefore

We, the members of the National Academy of Sciences, assembled at the autumn 1972 meeting, urge that textbooks of the sciences, utilized in the public schools of the nation, be limited to the exposition of scientific matter.

Passed by members of the National Academy of Sciences at the business session of the autumn meeting, 17 October 1972.

NATIONAL ACADEMY OF SCIENCES (1984)
Science and Creationism: A View from the National Academy of Sciences

State legislatures are considering, and some have passed, bills that would require the introduction of biblical creationism in science classes. Local school boards have passed ordinances to restrict the teaching of evolution or to require what is called a "balanced treatment" of creationism and evolution. Publishers of science textbooks are under pressure to deemphasize evolution while adding course material on "creation science."

The teaching of creationism as advocated by the leading proponents of "creation science" includes the following judgments: (1) the earth and universe are relatively young, perhaps only 6,000 to 10,000 years old; (2) the present form of the earth can be explained by "catastrophism," including a worldwide flood; and (3) all living things (including humans) were created miraculously, essentially in the forms we now find them. These teachings may be recognized as having been derived from the accounts of origins in the first two chapters of Genesis.

Generations of able and often devout scientists before us have sought evidence for these teachings without success. Others have given us hypotheses about the origin and history of the earth and the universe itself. These hypotheses have been tested and validated by many different lines of inquiry. With modifications to include new findings, they have become the central organizing theories that make the universe as a whole intelligible, lend coherence to all of science, and provide fruitful direction to modern research. The hypothesis of special creation has, over nearly two centuries, been repeatedly and sympathetically considered and rejected on evidential grounds by qualified observers and experimentalists. In the forms given in the first two chapters of Genesis, it is now an invalidated hypothesis. To reintroduce it into the public schools at this time as an element of science teaching would be akin to requiring the teaching of Ptolemaic astronomy or pre-Columbian geography.

Confronted by this challenge to the integrity and effectiveness of our national educational system and to the hard-won evidence-based foundations of science, the National Academy of Sciences cannot remain silent. To do so would be a dereliction of our responsibility to academic and intellectual freedom and to the fundamental principles of scientific thought. As a historic representative of the scientific profession and designated advisor to the Federal Government in matters of science, the Academy states unequivocally that the tenets of "creation science" are not supported by scientific evidence, that creationism has no place in a science curriculum at any level, that its proposed teaching would be impossible in any constructive sense for well-informed and conscientious science teachers, and that its teaching would be contrary to the nation's need for a scientifically literate citizenry and for a large, well-informed pool of scientific and technical personnel.

THE CENTRAL SCIENTIFIC ISSUES

Five central scientific issues are critical to consideration of the treatment in school curricula of the origin and evolution of the universe and of life on earth . . . .

THE NATURE OF SCIENCE

It is important to clarify the nature of science and to explain why creationism cannot be regarded as a scientific pursuit. The claim that equity demands balanced treatment of the two in the same classroom reflects misunderstanding of what science is and how it is conducted. Scientific investigators seek to understand natural phenomena by direct observation and experimentation. Scientific interpretations of facts are always provisional and must be testable. Statements made by any authority, revelation, or appeal to the supernatural are not germane to this process in the absence of supporting evidence. In creationism, however, both authority and revelation take precedence over evidence. The conclusions of creationism do not change, nor can they be validated when subjected to test by the methods of science. Thus, there are profound differences between the religious belief in special creation and the scientific explanations embodied in evolutionary theory. Neither benefits from the confusion that results when the two are presented as equivalent approaches in the same classroom. . . .

Special creation is neither a successful theory nor a testable hypothesis for the origin of the universe, the earth, or of life thereon. Creationism reverses the scientific process. It accepts as authoritative a conclusion seen as unalterable and then seeks to support that conclusion by whatever means possible.

In contrast, science accommodates, indeed welcomes, new discoveries: its theories change and its activities broaden as new facts come to light or new potentials are recognized. Examples of events changing scientific thought are legion. . . Prior acceptance of the fixed ad hoc hypothesis of creationism — ideas that are certified as untestable by their most ardent advocates — would have blocked important advances that have led to the great scientific achievements of recent years. Truly scientific understanding cannot be attained or even pursued effectively when explanations not derived from or tested by the scientific method are accepted.

SCIENTIFIC EVIDENCE ON THE ORIGIN OF THE UNIVERSE AND THE EARTH

The processes by which new galaxies, stars, and our own planetary system are formed are sometimes referred to as the "evolution" of the universe, the stars, and the solar system. The word evolution in this context has a very different meaning than it does when applied to the evolution of organisms.

Evidence that the evolution of the universe has taken place over at least several billion years is overwhelming. Among the most striking indications of this process are the receding velocities of distant galaxies. This general expansion of the universe was first noted in the late 1920s and early 1930s…astronomers today estimate that the expansion probably began some 10 to 20 billion years ago.

The invariant spontaneous decay of the radioactive isotopes of some elements provides further evidence that the universe is billions of years old. Analyses of the relative abundances of radioactive isotopes and their inert decay products in the earth, meteorites, and moon rocks all lead to the conclusion that these bodies are about 4.5 billion years old.…

A major reason for the creationists' opposition to the geological record and evolution is their belief that earth is relatively young, perhaps only a few thousand years old. In rejecting evidence for the great age of the universe, creationists are in conflict with data from astronomy, astrophysics, nuclear physics, geology, geochemistry, and geophysics. The creationists' conclusion that the earth is only a few thousand years old was originally reached from the timing of events in the Old Testament. . . .

...Although it was Darwin, above all others, who first marshaled the convincing critical evidence for biological evolution, earlier alert scholars recognized that the succession of living forms on the earth had changed systematically within the passage of geological time.

As applied to biology, a distinction is to be drawn between the questions (1) whether and (2) how biological evolution happened. The first refers to the finding, now supported by an overwhelming body of evidence, that descent with modification occurred during more than 2.7 billion years of earth's history. The second refers to the theory explaining how those changes developed along the observed lineages. The mechanisms are still undergoing investigation; the currently favored theory is an extensively modified version of Darwinian natural selection.

With that proviso we will now consider three aspects of biological evolution in more detail....

Relation by Common Descent: Evidence for relation by common descent has been provided by paleontology, comparative anatomy, biogeography, embryology, biochemistry, molecular genetics, and other biological disciplines. The idea first emerged from observations of systematic changes in the succession of fossil remains found in a sequence of layered rocks...

In Darwin's time, however, paleontology was still a rudimentary science, and large parts of the geological succession of stratified rocks were unknown or inadequately studied. Darwin, therefore, worried about the rarity of truly intermediate forms. Creationists have then and now seized on this as a weakness in evolutionary theory. Indeed, although gaps in the paleontological record remain even now, many have been filled by the researches of paleontologists since Darwin's time. Hundreds of thousands of fossil organisms found in well-dated rock sequences represent a succession of forms through time and manifest many evolutionary transitions.... There have been so many discoveries of intermediate forms between fish and amphibians, between amphibians and reptiles, between reptiles and mammals, and even along the primate line of descent that it is often difficult to identify categorically the line to which a particular genus or species belongs....

Although creationists claim that the entire geological record, with its orderly succession of fossils, is the product of a single universal flood that lasted a little longer than a year and covered the highest mountains to a depth of some 7 meters a few thousand years ago, there is clear evidence in the form of intertidal and terrestrial deposits that at no recorded time in the past has the entire planet been under water. The belief that all this sediment with its fossils was deposited in an orderly sequence in a year's time defies all geological observations and physical principles concerning sedimentation rates and possible quantities of suspended solid matter. We do not doubt that there were periods of unusually high rainfall or that extensive flooding of inhabited areas has occurred, but there is no scientific support for the hypothesis of a universal, mountain-topping flood.

Inferences about common descent derived from paleontology have been reinforced by comparative anatomy. The skeletons of humans, dogs, whales, and bats are strikingly similar, despite the different ways of life led by these animals and the diversity of environments in which they have flourished. The correspondence, bone by bone, can be observed in every part of the body, including the limbs. Yet a person writes, a dog runs, a whale swims, and a bat flies — with structures built of the same bones. Scientists call such structures homologous and have concurred that they are best explained by common descent.

Biogeography also has contributed evidence for common descent.... Creationists contend that the curious facts of biogeography result from the occurrence of a special creationary event. A scientific hypothesis proposes that biological diversity results from an evolutionary process whereby the descendants of local or migrant predecessors became adapted to their diverse environments. A testable corollary of that hypothesis is that present forms and local fossils should show homologous attributes indicating how one is derived from the other. Also, there should be evidence that forms without an established local ancestry had migrated into the locality. Whenever such tests have been carried out, these conditions have been confirmed.

Embryology, the study of biological development from the time of conception, is another source of independent evidence for common descent. Barnacles, for instance, are sedentary crustaceans with little apparent similarity to such other crustaceans as lobsters, shrimps, or copepods. Yet barnacles pass through a free-swimming larval stage, in which they look unmistakably like other crustacean larvae. The similarity of larval stages supports the conclusion that all crustaceans have homologous parts and a common ancestry....

Molecular Biology and the Degree of Relationship: Very recent studies in molecular biology have independently confirmed the judgments of paleontologists and classical biologists about relationships among lineages and the order in which species appeared within lineages. They have also provided detailed information about the mechanisms of biological evolution.

DNA, the hereditary material within all cells, and the proteins encoded by the genes in the DNA both offer extensive information about the ancestry of organisms. Analysis of such information has made it possible to reconstruct evolutionary events that were previously unknown, and to confirm and date events already surmised but not precisely dated.

In unveiling the universality of the chemical basis of heredity, molecular biology has profoundly affirmed common ancestry. In all organisms — bacteria, plants, and animals, including humans — the hereditary information is encoded in DNA, which is in all instances made up of the same four subunits called nucleotides. The genetic code by which the information contained in the nuclear DNA is used to form proteins is essentially the same in all organisms. Proteins in all organisms are invariably composed of the same 20 amino acids, all having a "left-handed" configuration, although there are amino acids in nature with both "right-" and "left-handed" configurations. The metabolic pathways through which the most diversified organisms produce energy and manufacture cell components are also essentially the same. This unity reveals the genetic continuity of living organisms, thereby giving independent confirmation of descent from a common ancestry. There is no other way consistent with the laws of nature and probability to account for such uniformity...

HUMAN EVOLUTION

Studies in evolutionary biology have led to the conclusion that mankind arose from ancestral primates. This association was hotly debated among scientists in Darwin's day, before molecular biology and the discovery of the now abundant connecting links. Today, however, there is no significant scientific doubt about the close evolutionary relationships among all primates or between apes and humans. The "missing links" that troubled Darwin and his followers are no longer missing. Today, not one but many such connecting links, intermediate between various branches of the primate family tree, have been found as fossils. These linking fossils are intermediate in form and occur in geological deposits of intermediate age. They thus document the time and rate at which primate and human evolution occurred...

THE ORIGIN OF LIFE

Scientific research on the origin of life is in an exploratory phase, and all its conclusions are tentative. We know that the organisms that lived on earth 2 billion or more years ago were simply microbial forms.... Experiments conducted under plausible primitive-earth conditions have resulted in the production of amino acids, large protein-like molecules made from long chains of amino acids, the nucleotide components of DNA, and DNA-like chains of these nucleotides. Many biologically interesting molecules have also been detected by astronomers using radiotelescopes. We can, therefore, explain how the early oxygen-free earth provided a hospitable site for the accumulation of molecules suitable for the construction of living systems.

For those who are studying aspects of the origin of life, the question no longer seems to be whether life could have originated by chemical processes involving nonbiological components but, rather, what pathway might have been followed. The data accumulated thus far imply selective processes. Prebiological chemical evolution is seen as a trial-and-error process leading to the success of one or more systems built from the many possible chemical components. The system that evolved with the capability of self-replication and mutation led to what we now define as a living system.

CONCLUSION

Scientists, like many others, are touched with awe at the order and complexity of nature. Religion provides one way for human beings to be comfortable with these marvels. However, the goal of science is to seek naturalistic explanations for phenomena — and the origins of life, the earth, and the universe are, to scientists, such phenomena — within the framework of natural laws and principles and the operational rule of testability.

It is, therefore, our unequivocal conclusion that creationism, with its account of the origin of life by supernatural means, is not science. It subordinates evidence to statements based on authority and revelation. Its documentation is almost entirely limited to the special publications of its advocates. And its central hypothesis is not subject to change in light of new data or demonstration of error. Moreover, when the evidence for creationism has been subjected to the tests of the scientific method, it has been found invalid.

No body of beliefs that has its origin in doctrinal material rather than scientific observation should be admissible as science in any science course. Incorporating the teaching of such doctrines into a science curriculum stifles the development of critical thinking patterns in the developing mind and seriously compromises the best interests of public education. This could eventually hamper the advancement of science and technology as students take their places as leaders of future generations.

Excerpts from "Science and Creationism: A View from the National Academy of Sciences," National Academy Press, Washington, DC 1984. Omissions of short phrases are not identified, but omissions of several sentences or more, usually of examples and argumentation in support of the central point, are indicated by ellipses. The editor has not made any additions.

New Orleans Geological Society

Science and Evolution vs Creationism and Louisiana Act 685 (1981): "Balanced Treatment for Evolution-Science and Creation-Science in Public School Instruction"

The New Orleans Geological Society, an organization of professional earth scientists, takes the position that science classes in Louisiana public schools should teach scientifically accurate and scientifically relevant material. The Society, therefore, disagrees with Louisiana Act 685 of 1981, the law for "Balanced Treatment of Creation-Science and Evolution-Science in Public School Instruction."

"Science" generally is defined as the systematic study of the activities of nature by accumulation of evidence that allows people to understand natural processes. A scientific theory is an idea, based upon a wealth of evidence, that describes and predicts conditions in nature. "Theory" — to a scientist — is a concept firmly grounded in and based upon facts, contrary to the popular conception that it is a hazy notion or undocumented hypothesis. Theories do not become facts; they explain facts. A theory must be verifiable; if evidence is found that contradicts the stated theory, the theory must be modified or discarded. In this manner, general knowledge is advanced. Scientific theories must provide new avenues for investigation and cannot be accepted on faith. Scientific facts supporting theories are presented to the scientific community in the form of published literature for examination by peers and by anyone else interested in the subject. In summary, science is not a belief system. It is simply a method for studying and accumulating knowledge about nature.

Louisiana Act 685 defines "creation-science" as "...the scientific evidences for creation and inferences from those scientific evidences." However, creation-science does not meet the foregoing rigorous standards. Creation-science data almost invariably are of questionable quality, obsolete, or taken out of context from the scientific literature. Even well-known creation scientists such as Duane Gish of the Institute for Creation Research have readily admitted that creation-science is not at all scientific.

Documentation refuting scientific creationism has been presented by the National Academy of Sciences, the Geological Society of America and by members of the American Association for the Advancement of Science and of the United States Geological Survey. Their findings and the findings of this Society are:

The bulk of creation-science literature is not devoted to the presentation of any positive evidence for creationism. Most of its material is an attempt to refute the evidence for the age of the Earth and organic evolution as documented by the geologic record and detailed biological studies, as if such a refutation would, by itself, leave creationism as the only logical alternative.

It is easily demonstrable that fossils are the remains of once living organisms that can be placed in a taxonomic hierarchy supporting evolution. It is also proved that strata of a given geological age contain certain fossil types that are of distinctive character and that over a wide geographical area occur in the same sequences. These are observable facts despite creationist claims that paleontological data do not support evolution.

The age of the Earth as determined by various methods including radiometric dating of meteorites and of the Earth's rocks is approximately 4.6 billion years. Creationist criticisms of that age are based upon misinterpretation of valid data and upon obsolete data. Creationists have failed to produce one single reliable dating technique that supports their idea of a young (6,000-year-old) Earth.

Creationists, in their charge that the "gaps" in the fossil record refute evolution, ignore the hundreds of identifiable transition species that have been catalogued. Concentrating their criticism only on vertebrate fossil finds, creationists neglect the detailed fossil record of invertebrates, microfauna, and microflora whose evolutionary change over time is well documented. That evolution has occurred is a documented fact, not disputed within the scientific community.

Creationist statistics "proving" that the origin of life from inanimate matter is impossible are inaccurate. Such statistical calculations do not take into account laboratory evidence showing that organic matter does organize itself, and that organic molecules can carry on processes similar to life-sustaining biochemical actions outside the cell. Also omitted are astronomical observations that demonstrate the ubiquitous nature of organic matter throughout the solar system and the galaxy.

Arguments stating that thermodynamics precludes the evolution of life because evolution would run against the trend of order to disorder in nature misrepresent the science of thermodynamics. Such arguments are not based on any mathematical calculations. Thermodynamics does in fact show that entropy reversals can and do occur in a biological system that is open with respect to energy input, which is the case for the biosphere of the Earth.

Creationism, as a scientific concept, was dismissed over a century ago and subsequent research has only confirmed that conclusion. Scientific creationism threatens to do great damage to the credibility of legitimate scientific research and to data accumulated from the many varied and unrelated scientific disciplines that independently support organic evolution as a verifiable scientific concept because of its misuse of those data.

The Society, as stated in the introduction to this document, is against the teaching of creationism in our public schools as science along with evolution on an equal basis. The creationist concept of "equal time" has no place in the advancement of science. If an idea can be shown to have no scientific merit, it must either be modified in light of available facts or new data or discarded regardless of how much its proponents believe in it. Creationism is such an idea. It is based on a preconceived notion, not upon any observations of nature and the world around us. The Society has no objection to people wanting to believe that the universe, the Earth, and its residents were created in 6 days, 6,000 years ago. However, those people must realize that such ideas are religious in nature and cannot be called scientific.

By advocating this position, the Society is not taking a stand against any particular religious belief. Science and religion are two different disciplines that are not in conflict with one another. Science is not atheistic; it is non-theistic, and it makes no judgment of religion. The Society feels that religious views have no place in the science classroom.

At the same time, the Society supports the teaching of evolution in science classes precisely because it is legitimate science. As a nation, we live in a society heavily influenced by science and technology. Evolution is a basic scientific concept. People do not have to "believe" in it, but they should understand evolution and how and why it came about.

It is because the system of scientific education in this country has declined in recent years that laws such as Act 685 became possible. Legislation such as this Act, that attempts to legislate what should be taught as science in public schools, ignores one simple fact: scientific findings cannot be altered by public opinion. It is irrelevant that some public opinion polls show approval of creationism being taught alongside evolution. Laws that require non-scientific ideas such as creationism to be taught as current scientific thought alongside established scientific principles such as evolution, or teach neither, do not promote free inquiry — they stifle it. Scientific research and education cannot take place in such a coercive atmosphere.

1985

NEW YORK ACADEMY OF SCIENCES

Mandating the study of scientific creationism in the public schools of New York State, as embodied in New York State Assembly Bill 8569 and New York State Senate Bill 8473, by legislative mandate is viewed by the New York Academy of Sciences as an attempt to introduce, by fiat, religious dogma into an arena where verifiability is paramount to the subject matter. It would constitute a very serious breach of the concept of the separation of Church and State. Scientific Creationism is a religious concept masquerading as a scientific one.

Science attempts to explain the physical world through verifiable and repeatable data. Through its rigorous application of inductive and deductive logic, science asks how physical phenomena occur. It attempts to explain the processes that bring about the phenomena that exist now or have existed in the past.

The concept of evolution in biology is an attempt to ascertain how life may have originated, developed and diversified on the planet Earth. Concepts such as that of evolution are developed within the framework of natural laws. The methodology of science aims to ascertain these laws from experimental data. Science accepts the theories or hypotheses that best "fit" these data.

Science modifies established theories in the light of new experimental data. It is receptive to new theories, if they withstand the tests of scientific methodology.

The concept of evolution is incorporated within many scientific disciplines. Scientific data supplied from these many disciplines have contributed to a more thorough understanding of the mechanism of evolution. The theory itself does not rest on any single branch of science.

Because of inherently different methodologies of science and of religion, there is no overlapping area where the methods of science can be applied to religion or vice versa. There is no way for science to test the various accounts of creation held by the world's religions. These accounts depend upon the acceptance of supernatural phenomena and are not subject to scientific investigation. Their proponents demand that these accounts be accepted on faith, and are properly the province of religion. The methodologies of science cannot be used for their evaluation.

The subject known as "Scientific Creationism" is lacking in scientific substance; we reject it for inclusion in science curricula.

For these reasons, the New York Academy of Sciences strongly opposes the introduction of "Scientific Creationism" into any science curricula of the public schools of New York State.

Passed by the Board of Governors of the New York Academy of Sciences on 22 May 1980.

North Carolina
Academy of Science

Intellectual freedom and the quality of science education in North Carolina, and the competency of future generations of North Carolinians to make wise decisions concerning science and technology, are being threatened by groups pressuring educators to present creationism as a scientifically viable alternative to evolution. Textbooks are being censored; authors, science teachers, and school boards are being intimidated; and science curricula are being modified in ways that accommodate non-scientific points of view and reject principles accepted by the scientific community.

The North Carolina Academy of Science strongly opposes any measure requiring or coercing public school educators either to include creationism in science curricula or to limit the inclusion of evolution in those same curricula. Principles and concepts of biological evolution are basic to the understanding of science. Students who are not taught these principles, or who hear creationism presented as a scientific alternative to them, will not be receiving an education based on modern scientific knowledge. Their ignorance about evolution will seriously undermine their understanding of the world and the natural laws governing it, and their introduction to creationism as "scientific" will give them false ideas about scientific methods and criteria. Yet we must give students who will face the problems of the 20th and 21st centuries the best possible education.

Creationists claim that biological evolution is a religious tenet; in fact it is one of the cornerstones of modern science. More than 50 years ago the North Carolina Academy of Science adopted a resolution declaring evolution an established law of nature, and since then extensive data have accumulated which further reinforce the confidence of the scientific community in the validity of evolution and help clarify the mechanisms through which evolution operates. Scientists agree that organisms now living on the earth are derived from pre-existing organisms which, over long periods of time measured in billions of years, have changed from the simplest ancestors to the diverse and complex biota now in existence. Scientists further agree that there was a time when the earth was devoid of life, and that life developed through natural processes. The evidences supporting these conclusions are extensive, are drawn from many disciplines of science, and are mutually corroborative. They have withstood tests and searching criticism as rigorous as that to which any scientific principles have been subjected. No scientific hypothesis suggested as an alternative to evolution has succeeded in explaining relevant natural phenomena. Moreover, insights provided by evolutionary principles have been the basis for progress in the biological and biomedical sciences which has benefited mankind in many ways.

There are important questions remaining, of course, about how evolution operates. We have made progress in this area during the past century, but debates about evolutionary mechanisms still go on today. Some creationists, in an attempt to discredit the principles of evolution, have emphasized these disagreements between scientists about how evolution takes place. But such discussion is a normal part of how science works; fruitful controversy plays an important role in stimulating scientific investigation and furthering scientific knowledge. Debate about evolutionary mechanisms in no way undermines scientists' confidence in the reality of evolution, any more than disagreement about the behavior of subatomic particles would lead scientists to doubt the existence of atoms.

Creationists contend that creationism is a scientific theory and therefore a valid alternative to evolution. But to quote from a statement by the National Science Teachers Association, "The true test of a theory in science is threefold: (1) its ability to explain what has been observed; (2) its ability to predict what has not been observed; and (3) its ability to be tested by further experimentation and to be modified by the acquisition of new data." Viewed in the context of these criteria, creationism is not scientific. There should be opportunity for full discussion of such non-scientific ideas in appropriate forums, but they have no place in science classes. The content of science courses must meet scientific criteria; to require equal time for discussion of non-science topics would destroy the integrity of science education.

Therefore, we the members of the North Carolina Academy of Science declare the following to be the position of the Academy on this issue:

The North Carolina Academy of Science strongly opposes the mandated inclusion of creationist views of origins in public school science classes. Furthermore, the Academy is strongly opposed to any mandated exclusion of the principles of evolution from public school instruction. We totally reject the concept, put forth by certain pressure groups, that evolution is itself a tenet of religion. And we assert that evolution is the only strictly scientific explanation for changes in the biota of the earth over time and for the existence and diversity of living organisms.

January 1982.

OHIO ACADEMY OF SCIENCE
Forced Teaching of Creationist Beliefs in Public School Science Education

W*hereas*, it is a responsibility of the Ohio Academy of Science to preserve the integrity of science; and

Whereas, science is a systematic method of investigation based on continuous experimentation, observation, and measurement leading to evolving explanations of natural phenomena, explanations which are continuously open to further testing; and

Whereas, evolution fully satisfies these criteria, irrespective of remaining debates concerning its detailed mechanisms; and

Whereas, the Academy respects the right of people to hold diverse beliefs about creation that do not come within the definitions of science; and

Whereas, Creationist groups are imposing beliefs disguised as science upon teachers and students to the detriment and distortion of public education in the United States;

Therefore, be it resolved that because "Creationist Science" has no scientific validity it should not be taught as science, and further, that the OAS views legislation requiring "Creationist Science" to be taught in public schools as a real and present threat to the integrity of education and the teaching of science; and

Be it further resolved that the OAS urges citizens, educational authorities, and legislators to oppose the compulsory inclusion in science education curricula of beliefs that are not amenable to the process of scrutiny, testing, and revision that is indispensable to science.

This resolution, identical to the AAAS resolution published two months earlier, was adopted by the Council of the OAS on 23 April 1982 and published in the Ohio Journal of Science 82(3):inside back cover, 1982.

OKLAHOMA ACADEMY OF SCIENCES

The scientific content of science courses should be determined by scientists and science educators and not by political directives. In particular, science teachers should not be required to teach, as science, ideas, models and theories that are clearly extra-scientific. An extra-scientific hypothesis, as such, might legitimately be discussed in a science class when examination of its logical construction and criteria for acceptance would illuminate the corresponding features of a scientific hypothesis and scientific method. Any requirement for equal time for such hypotheses is not justifiable.

Scientific hypotheses have a number of distinguishing properties, the foremost of which is that one should be able to deduce, from the basic postulates, logical consequences that can be tested against observation. Attention should be paid to the possible kinds of evidence that would falsify the hypothesis, rather than just the evidence that might confirm it. Other properties include:

1. The hypothesis should have more general consequences than those observations which initially suggested it. Thus it should be independently testable and not ad hoc.

2. It should be fruitful, suggesting new lines of research to pursue, raise new questions to by investigated by future research.

3. It should be logically consistent.

4. It should be consistent with the general scientific philosophy that the observed phenomena of the universe are real and that nature is consistent and understandable, that is, describable and explainable in terms of laws and theories.

Hypotheses that postulate miracles or supernatural events are falsified scientifically because they explicitly admit they cannot explain phenomena within their sphere of application. Furthermore, they are extra-scientific and non-explanatory because those phenomena are declared to be beyond human understanding. Thus they can not be considered alternate explanation to any scientific hypothesis because, by their very nature, they are anti-explanatory, seeking only to establish and perpetuate a mystery or mysteries.

All such hypotheses, models and theories that claim to be scientific should be required to meet the same criteria as do those hypotheses commonly considered to be scientific by the scientific community at large.

Adopted November 13, 1981
(earlier adopted by Oklahoma Science Teachers' Association)

SIGMA XI,
LOUISIANA STATE UNIVERSITY CHAPTER

The LSU Chapter of Sigma Xi urges the reconsideration and repeal of the "Balanced Treatment for Creation-Science and Evolution-Science Act" which in 1981 became part of Louisiana law.

The current science curriculum is the result of numerous discoveries and critical studies by scientists over many decades. The scientific process affords equal treatment to every theory by requiring it to face the evidence successfully before it becomes part of the science curriculum. The theory called "creation science" cannot successfully face the evidence. The Act constitutes intervention by the State to give that theory a standing it has not earned. The Act, if put into effect, would violate academic freedom and weaken science education. This is a time for strengthening educational standards and programs, particularly in science.

Approved by mail ballot of the membership and released 15 February 1982.

SOCIETY FOR AMATEUR SCIENTISTS

The Society for Amateur Scientists was founded to place the power, process, and promise of science within reach of everyone. SAS links science enthusiasts of all backgrounds and interests with world-class professional scientists, to empower amateurs to take part in the great scientific debates of our time as full members of the scientific community. Our mission is two-fold: to advance science by bringing untapped talent into the field, and to help create a more scientifically literate public.

The debate about teaching evolution and scientific creationism in the public schools has raged for decades. Is it appropriate for a grass roots science organization like ours to comment on this debate? Absolutely. The Society for Amateur Scientists was founded to educate people about how science works, what science tells us about our world, and how everyday people can take an active part in fascinating scientific issues. Some participants in this debate constantly distort science and misinform the public. Correcting misunderstandings is clearly part of any educational mission.

But there is a deeper concern. Our democracy depends on an informed and educated electorate. As science literacy suffers, so does our country. This is truer today then ever before as the voting public is faced with ever more technical issues about which they are asked to make informed choices. By not opposing bad science whenever we can, SAS would be implicitly aiding the forces of unreason to distort fundamental principles of science in the public mind. We believe that it is vital that all scientific organizations, including SAS, stand against bad science.

In the last 100 years, science has forged a profound understanding of many different fields which bear on the question of our origin. Genetics, astronomy, geology, paleontology, biology, physiology, anatomy and physics all speak with one voice. The universe is ancient, perhaps 15 billion years old. The earth too is ancient, perhaps 5 billion years old. And life is ancient, perhaps 2 billion years old.

The evidence is abundant and irrefutable. Life has changed drastically over earth's history. Since the first complex multi-cellular forms appeared about 650 million years ago organisms have lived, died and adapted to their environments through many violent upheavals on the planet. The one constant has been the process of change itself – of mutation and natural selection, the hammer and anvil by which nature has sculpted her handiwork into the imperfectly beautiful and intricate web of life that now covers the planet.

On the question of humanity, the data support only one conclusion – humans arose like all other beings with which we share the earth; through the random mutations altering our ancestors' bodies over eons, and natural selection blindly and mercilessly cutting away the chaff. Evolution is the great shaper of all life on earth.

Today, evolution is *the* unifying principle of biology. Nothing makes sense without it. True, it remains a very active field of research and many subtle and fascinating questions remain to be answered. However, that life has adapted and changed through time is as well established as the fact that the earth goes round the sun.

Evolution is science, and as such belongs in science classrooms. By contrast scientific creationism just doesn't make the grade. None of the arguments which scientific creationists make against evolution withstand scrutiny and most were first refuted nearly a century ago. And the creationists have never been able to marshal quality evidence that strongly supports their ideas.

This statement was approved by our Board of Directors. Amateur scientists are often fiercely independent, and some of our members do not accept evolution. While the Board of Directors respects their views and values their input, we wish to make it clear that SAS will never participate in creationist research. However, we do not restrict our membership to avowed evolutionists. As a scientific organization, we insist only that our members be willing to consider any position that can be supported by empirical evidence. In this we are quite unlike the Institute for Creation Research (ICR), the primary promoter of Scientific Creationism in public school, which requires its members to sign a statement attesting to their belief in the literal truth of the Bible. ICR's agenda is religion concealed in the guise of science. Their materials in particular have no place in a science classroom.

1994
Shawn Carlson, Ph.D.
Elizabeth Arsem
Paul MacCready, Ph.D.
Glenn T. Seaborg, Ph.D.

SOCIETY FOR THE STUDY OF EVOLUTION

I n 1952, Ernst Mayr stated that "the aims of the Society [for the Study of Evolution], through its journal and otherwise, reflect the conviction that the evolutionary approach will clarify many unsolved biological problems and will provide common goals and mutual comprehension among all the life sciences." The history of evolutionary studies has as its basis empirical documentation of bio-geographical distribution of species. Contributing to its development are rigorous horticultural and agricultural programs that have led to substantial improvements in world food supplies. More recently, evolutionary studies have been applied to conservation and to health-related fields such as disease epidemiology. Increasingly, evolutionary studies have been applied to conservation and to health related fields such as disease epidemiology. Increasingly, evolutionary studies are used to predict how the biological world responds to changing environments – environments that indisputably have changed over time. Evolutionary studies supply scientific explanations for past and present biological processes, based on currently observed biological processes. They have directly provided information, techniques, and even products that contribute to the improvement of human conditions and ecological welfare.

The study of evolution is an empirically based science which employs the scientific process of hypothesis testing. Hypotheses are either accepted or rejected, depending on the empirical evidence. The Society for the Study of Evolution employs a rigorous critical review process to ensure that these procedures are followed – that the empirical data support the conclusions – before a study is accepted as scientific. No hypothesis that cannot be tested empirically is acceptable as scientific to the Society. "Scientific creationism" cannot be empirically refuted. Rather, it has as its basis the unquestioned authority of a literal interpretation of religious texts. "Scientific creationism" does not employ hypothesis testing, does not use unbiased empirical data to support or refute hypotheses, and it has no scientific review process. It therefore cannot be considered to be scientific by the Society. The attitude that "scientific creationism" is an alternative hypothesis to evolution is scientifically untenable. Its inclusion in state-sponsored school curricula as a scientifically based hypothesis rather than as a religious faith is not acceptable.

The Society for the Study of Evolution maintains that evolutionary studies should be promoted in schools as a scientific approach to explaining biological phenomena – one that has contributed much to biotechnological advances, and one which has the potential to solve important problems in the physical relationship of human beings to the rest of the biological world.

Society of Vertebrate Paleontology (1986)

*B*e it resolved, that the Society of Vertebrate Paleontology opposes the teaching of so-called "creation science" or "scientific creationism" as a viable alternative to evolutionary explanations of the origin and history of the earth and of life, on the grounds that "creation science" or "scientific creationism" is in its essentials a body of religious doctrines rather than an embodiment of scientific process.

Be it further resolved, that the officers of the Society of Vertebrate Paleontology are hereby authorized to investigate the feasibility of associating the Society with one of the briefs of amicus curiae in the Louisiana creationism case now pending before the United States Supreme Court; and that, if feasible, the Society of Vertebrate Paleontology formally associate itself with such a brief opposing the teaching of "scientific creationism" as science.

Unanimously passed at the general business meeting held during the 46th annual meeting in Philadelphia, on 7 November 1986, and distributed by letter over the signature of SVP President Bruce J. MacFadden.

SOCIETY OF VERTEBRATE
PALEONTOLOGY (1994)

The fossil record of vertebrates unequivocally supports the hypothesis that vertebrates have evolved through time, from their first records in the early Paleozoic Era about 500 million years ago to the great diversity we see in the world today. The hypothesis has been strengthened by so many independent observations of fossil sequences that it has come to be regarded as a confirmed fact, as certain as the drift of continents through time or the lawful operation of gravity.

Paleontology relies for its evidence on two different but historically related fields, biology and geology. Evolution is the central organizing principle of biology, understood as descent with modification. Evolution is equally basic to geology, because the patterns of rock formations, geomorphology, and fossil distributions in the world make no sense without the underlying process of change through time. Sometimes this change has been gradual, and sometimes it has been characterized by violent upheaval. These processes can be seen on the Earth today in the forms of earthquakes, volcanoes, and other tectonic phenomena. Vertebrates have also evolved at a variety of rates, some apparently gradual, and some apparently rapidly. Although the fossil record is not complete, and our knowledge of evolution will always be less than entire, the evidence for the progressive replacement of fossil forms has been adequate to support the theory of evolution for over 150 years, well before genetic mechanisms of evolutionary change were understood. Paleontologists may dispute, on the basis of the available evidence, the tempo and mode of evolution in a particular group at a particular time, but they do not argue about whether evolution took place: that is a fact.

The fossil record has long been seen as a search for "ancestors" of living forms and of other fossil forms. Some fossil vertebrates appear to have no features that debar them from ancestry to other groups, and so could be seen as potential ancestors. Nevertheless, paleontologists do not focus on a search for direct ancestors, but rather look for sets of evolutionarily derived characters that are shared by fossil taxa that can then be linked as each other's closest known relatives. Proceeding in this way, paleontologists have clarified in recent years a great many mysteries about the origins and interrelationships of major groups of vertebrates, including birds, dinosaurs and their relatives, lizards and snakes, Mesozoic marine reptiles, turtles, mammals and their relatives, amphibians, the first tetrapods, and many groups of fishes. At the same time, techniques of geologic dating, including magnetostratigraphy, radiometric dating of many different isotopes of common elements, lithostratigraphy, and biostratigraphy, have provided independent lines of evidence for determining age relationships of the sediments in which fossils are found. This evidence from the principles and techniques of chemistry and physics support the finds of paleontology based on paleobiological and geological analyses, making the theory of evolution the only robust scientific explanation for the patterns of life on Earth.

Evolution is fundamental to the teaching of good biology and geology, and the vertebrate fossil record is an excellent set of examples of the patterns and processes of evolution through time. We therefore urge the teaching of evolution as the only possible reflection of our science. Any attempt to compromise the patterns and processes of evolution in science education, to treat them as less than robust explanations, or to admit "alternative" explanations not relying upon sound evolutionary observations and theory, misrepresents the state of our science and does a disservice to the public. Textbooks and other instructional materials should not indulge in such misrepresentation, educators should shun such materials for classroom use, and teachers should not be harassed or impeded from teaching vertebrate evolution as it is understood by its practitioners. The record of vertebrate evolution is exciting, inspirational, instructive, and enjoyable, and it is our view that everyone should have the opportunity and the privilege to understand it as paleontologists do.

Adopted November, 1994

SOUTHERN ANTHROPOLOGICAL SOCIETY

The Southern Anthropological Society deplores the intrusion of a particular religious doctrine into public school classrooms under the guise of so-called "scientific creationism."

These doctrines claim that a literalist reading of the account of the origins of the earth and life on it, as contained in the initial chapters of the book of Genesis, is supported by acceptable scientific evidence.

This interpretation treats a religious text as a scientific theory, which would seem to misrepresent both religion and science. The overwhelming evidence of the sciences – cosmology, geology, biology, anthropology, among others – indicates that the earth and all living forms on it have evolved from a simpler state, although, as in all ongoing science, theories as to how this took place continue to be revised in detail.

There is no necessary conflict between religious belief and inquiry into the natural world.

The institutionalization of creationist doctrine in the school curriculum will lead to the crippling of scientific inquiry as well as to the blurring of the important constitutional distinction between church and state.

Passed at the general business meeting of the Southern Anthropological Society on 16 April 1982 and published in The Southern Anthropologist *(SAS newsletter), 10(1):1,7.*

WEST VIRGINIA ACADEMY OF SCIENCE

*B*e it resolved that the West Virginia Academy of Science adopts the following position statement on the relation between science and religion, and on their places in science classrooms in public schools.

In the modern world, science is one important way of organizing human experience. That there are other important ways is evident from the existence of diverse religions and other nonscientific systems of thought.

Our nation requires well trained scientists and scientifically literate citizens who understand the values and limitations of science. Therefore, science courses should not only convey the important conclusions of modern science, but should also help students to understand the nature of scientific thought, and how it differs from other modes of thought.

Teachers are professionally obligated to treat all questions as objectively as possible. Questions regarding the relation between science and various religions may arise. To the extent that a teacher feels competent to do so, he or she should be free to respond to such questions. It is appropriate to show why science limits itself to ways of reasoning that can only produce naturalistic explanations. However, teachers and students should be free to challenge the presuppositions of science and to question their adequacy as a basis for a religion or world view. Ideas offered seriously by students deserve a serious response. They will never be ridiculed by teachers with high professional standards. Furthermore, teachers should make it clear that students will be evaluated on their understanding of the concepts studied, and not on their personal beliefs regarding those concepts.

Dogmatic assertions are inconsistent with objective consideration of any subject. Science is always tentative and does not pretend to offer ultimate truth. Nevertheless, there is an overwhelming consensus among scientists that the earth is several billion years old, that living organisms are related by descent from common ancestors, and that interpretation of all available evidence by scientific standards renders contrary claims highly implausible.

"Scientific creationism," which does challenge these conclusions, is a point of view held only by those who insist that the principle of biblical inerrancy and perspicuity must take precedence over all scientific considerations. This viewpoint is religious. Their claim that scientific creationism is independent of biblical creationism, which they admit is religious, is demonstrably false. The consistently poor scholarship of their attempts to defend scientific creationism suggests that their dominating principle can be accepted on faith but is not compatible with scientific standards of reasoning. It is clear that scientific creationism and science are two distinct systems of thought. It should be noted that other religions, including other varieties of Christianity, are also distinct from science, but are compatible with it.

Scientific creationists have defined the issue in such a way that their point of view on one side is contrasted with all other points of view lumped together on the other side, even though some of these other points of view also consider themselves creationist. Their demand that public schools devote equal time and resources to scientific creationism is in effect a demand that their religion be accorded special status and that schools purchase large quantities of books from their publishing houses, even though these books demonstrably represent poor scholarship. It is an attempt to win by legislative decree what they have been unable to win through scholarly argument. Proposals for equal-time legislation are unwise.

Be it resolved that the West Virginia Academy of Science endorses and adopts the AAAS (American Association for the Advancement of Science) resolution on Forced Teaching of Creationist Beliefs in Public School Science Education. This resolution, adopted by the AAAS Board of Directors and AAAS Council in January, 1982, read as follows: [*See the 1982 statement by the American Association for the Advancement of Science on page 25.*]

Passed at the WVAS annual business meeting on 3 April 1982 and published in the Proceedings of the West Virginia Academy of Science, *54:154-155.*

PART THREE

Religious Organizations

AMERICAN JEWISH CONGRESS

The American Jewish Congress is a national organization committed to the vigorous enforcement of the First Amendment provision requiring separation of church and state. The First Amendment provides "Congress shall make no law respecting an establishment of religion." This provision — often called the establishment clause – forbids the government from performing or aiding in the performance of a religious function.

Our appearance at this hearing today arises from our concern that Proclamation 60 (both alone and together with Board Rule 5) abrogates the establishment clause in three fundamental ways. The first constitutional deficiency lies in the Proclamation's glaring omission of any reference to the Darwinian theory of evolution. The second constitutional deficiency lies in the Board Rule's requirement that evolution be singled out for a special negative treatment not required in connection with the teaching of any other scientific theory. The third constitutional deficiency arises from the fact that the proposed textbook standards allow for the teaching of scientific creationism. Despite attempts to describe scientific creationism as scientific theory, it is our position that scientific creationism is a religious theory and that, therefore, the First Amendment's establishment clause prohibits its being taught as science in public school classes.

It seems apparent that, in establishing the proposed textbook standards, the intent of the State Board of Education has been to avoid conflict with a particular religious doctrine and to allow for the inclusion of religious theory in the science curriculum. The United States Supreme Court has made clear that the approach employed by Proclamation 60 is unconstitutional. In 1968, in a case titled *Epperson vs Arkansas*, an Arkansas biology teacher asked the Supreme Court to declare void a state statute which prohibited the teaching of evolution and which prohibited the selection, adoption or use of textbooks teaching that doctrine. The Supreme Court held that the statute was unconstitutional. In its opinion the Supreme Court stated:

"The First Amendment's prohibition is absolute. It forbids alike the preference of a religious doctrine or the prohibition of a theory which is deemed antagonistic to a particular dogma."

Under the standards so clearly articulated by the Supreme Court, Proclamation 60 and Board Rule 5, as presently written, fail to satisfy the constitutional requirement of separation of church and state. In order to comply with the applicable constitutional provisions, the proclamation and board rule should be revised in three ways. First, evolution should be clearly included in the science curriculum. Second, evolution should be taught as are all scientific theories and should not be singled out for special negative comment. Finally, the proposed textbook standards should make clear that scientific creationism is not to be taught as scientific theory. Rather, because there is no constitutional objection to teaching about religion, public school teachers should simply tell their students, when evolution is taught, that there are certain religious groups whose members do not accept the Darwinian theory and advise them to consult with their parents or religious advisors for further guidance on the subject.

The American Jewish Congress believes that this approach is not only fully consistent with the Constitution but is also an effective means by which to resolve objections to the teaching of evolution.

Should the Board of Education fail to take the steps necessary to make the Proclamation constitutional, then the result could lead to textbooks which do not meet constitutional standards. And that mistake would be a costly one to the taxpayers.

Testimony in behalf of the American Jewish Congress by spokesperson Nina Cortell before the Texas State Board of Education, responding to Proclamation 60, setting forth specific content rules for biology and science textbooks to be adopted in 1984.

AMERICAN SCIENTIFIC AFFILIATION:
A Voice for Evolution As Science

After polling the membership on its views, the *Executive Council of the American Scientific Affiliation* hereby directs the following Resolution to public school teachers, administrators, school boards, and producers of elementary and secondary science textbooks or other educational materials:

Because it is our common desire to promote excellence and integrity in science education as well as in science; and

Because it is our common desire to bring to an end wasteful controversy generated by inappropriate entanglement of the scientific concept of evolution with political, philosophical, or religious perspectives;

We strongly urge that, in science education, the terms evolution and theory of evolution should be carefully defined and used in a consistently scientific manner; and

We further urge that, to make classroom instruction more stimulating while guarding it against the intrusion of extra-scientific beliefs, the teaching of any scientific subject, including evolutionary biology, should include 1) forceful presentation of well-established scientific data and conclusions; 2) clear distinction between evidence and inference; and 3) candid discussion of unsolved problems and open questions.

Adopted by the Executive Council of the American Scientific Affiliation on December 7, 1991. ASA was founded in 1941 as a nationwide fellowship of evangelical Christians trained in science. Its vision is "To have science and theology interacting and affecting one another in a positive light." The 1991 resolution was preceded by a background statement citing various definitions of evolution and identifying "scientific creationism" at one extreme and "evolutionary naturalism" at the other as "essentially religious doctrine masquerading as science." First published in ASA's journal, Perspectives on Science & Christian Faith *(Vol. 44, No. 4, p. 252, Dec. 1992), the resolution and its background statement also appear in the 1993 edition of* Teaching Science in a Climate of Controversy, *a guidebook for high school teachers from ASA, P.O. Box 668, Ipswich, MA 01938.*

CENTER FOR THEOLOGY
AND THE NATURAL SCIENCES

The universe is more mysterious than either science or religion can ever fully disclose, and the urgencies of humankind and the natural environment demand an honest interaction between the discoveries of nature, the empowerment afforded us by appropriate technology, the inherent value of the environment, and the demand that we commit ourselves to a future in which all species can flourish. We can no longer afford the stalemate of past centuries between theology and science, for this leaves nature Godless and religion worldless. When this happens, our culture, hungering after science for something to fill the void of its lost spiritual resources, is easy prey to New Age illusions wrapped in scientific-sounding language — the 'cosmic self-realization movement' and the 'wow of physics' — while our 'denatured' religion, attempting to correct social wrong and to provide meaning and support for life's journey, is incapable of making its moral claims persuasive or its spiritual comfort effective because its cognitive claims are not credible. Nor can we allow science and religion to be seen as adversaries, for they will be locked in a conflict of mutual conquest, such as "creation science" which costs religion its credibility or a philosophical stance of "scientific materialism" which costs science its innocence....

Excerpted from the Mission Statement of the Center for Theology and the Natural Sciences, Berkeley, California

CENTRAL CONFERENCE
OF AMERICAN RABBIS

On Creationism in School Textbooks

Whereas the principles and concepts of biological evolution are basic to understanding science; and

Whereas students who are not taught these principles, or who hear "creationism" presented as a scientific alternative, will not be receiving an education based on modern scientific knowledge; and

Whereas these students' ignorance about evolution will seriously undermine their understanding of the world and the natural laws governing it, and their introduction to other explanations described as "scientific" will give them false ideas about scientific methods and criteria,

Therefore be it resolved that the Central Conference of American Rabbis commend the Texas State Board of Education for affirming the constitutional separation of Church and State, and the principle that no group, no matter how large or small, may use the organs of government, of which the public schools are among the most conspicuous and influential, to foist its religious beliefs on others;

Be it further resolved that we call upon publishers of science textbooks to reject those texts that clearly distort the integrity of science and to treat other explanations of human origins for just what they are — beyond the realm of science;

Be it further resolved that we call upon science teachers and local school authorities in all states to demand quality textbooks that are based on modern, scientific knowledge and that exclude "scientific" creationism;

Be it further resolved that we call upon parents and other citizens concerned about the quality of science education in the public schools to urge their Boards of Education, publishers, and science teachers to implement these needed reforms.

Adopted at the 95th Annual Convention of the Central Conference of American Rabbis, 18-21 June 1984, at Grossinger's, New York.

PASTORAL LETTER,
The Rt. Rev. Bennett J. Sims, Episcopal Bishop of Atlanta
A Pastoral Statement on Creation and Evolution

G race to you and peace from God our Father and from the Lord Jesus Christ.

Legislation is pending before the Georgia State Legislature which calls for the public financing and teaching of Scientific Creationism as a counter-understanding to Evolution, wherever the evolutionary view is taught in the public schools.

Scientific Creationism understands the cosmos and the world to have originated as the Bible describes the process in the opening chapters of Genesis.

The 74th Annual Council of the Diocese of Atlanta, in formal action on January 31, 1981, acted without a dissenting vote to oppose by resolution any action by the Georgia Legislature to impose the teaching of Scientific Creationism on the public school system. A copy of the resolution is attached to this Pastoral.

It seems important that the Episcopal Church in this diocese add to its brief resolution a statement of its own teaching. The office of Bishop is historically a teaching office, and I believe it is timely to offer instruction as to this Church's understanding of what has become a contested public issue.

To begin with creation is a fact. The world exists. We exist. Evolution is a theory. As a theory, evolution expresses human response to the fact of creation, since existence raises questions: how did creation come to be, and why?

The question of why is the deeper one. It takes us into the realm of value and purpose. This urgent inquiry is expressed in human history through religion and statements of faith. Christians cherish the Bible as the source book of appropriating the point and purpose of life. We regard the Bible as the Word of God, His revelation of Himself, the meaning of His work and the place of humanity in it.

The question of how is secondary, because human life has been lived heroically and to high purpose with the most primitive knowledge of the how of creation. Exploration of this secondary question is the work of science. Despite enormous scientific achievement, humanity continues to live with large uncertainty. Science, advancing on the question of how, will always raise as many questions as it answers. The stars of the exterior heavens beyond us and the subatomic structure of the interior deep beneath us beckon research as never before.

Religion and science are therefore distinguishable, but in some sense inseparable, because each is an enterprise, more or less, of every human being who asks why and how in dealing with existence. Religion and science interrelate as land and water, which are clearly not the same but need each other, since the land is the basin for all the waters of the earth and yet without the waters the land would be barren of the life inherent to its soil.

In the Bible the intermingling of why and how is evident, especially in the opening chapters of Genesis. There the majestic statements of God's action, its value and the place of humanity in it, use an orderly and sequential statement of method. The why of the divine work is carried in a primitive description of how the work was done.

But even here the distinction between religion and science is clear. In Genesis there is not one creation statement but two. They agree as to why and who, but are quite different as to how and when. The statements are set forth in tandem, chapter one of Genesis using one description of method and chapter two another. According to the first, humanity was created, male and female, after the creation of plants and animals. According to the second, man was created first, then the trees, the animals and finally the woman and not from the earth as in the first account, but from the rib of the man. Textual research shows that these two accounts are from two distinct eras, the first later in history, the second earlier.

From this evidence, internal to the very text of the Bible, we draw two conclusions.

First, God's revelation of purpose is the overarching constant. The creation is not accidental, aimless, devoid of feeling. Creation is the work of an orderly, purposeful Goodness. Beneath and around the cosmos are the everlasting arms. Touching the cosmos at every point of its advance, in depth and height, is a sovereign beauty and tenderness. Humanity is brooded over by an invincible Love that values the whole of the world as very good; that is the first deduction: God is constant.

Second, creation itself and the human factors are inconstant. Creation moves and changes. Human understanding moves and changes. Evolution as a contemporary description of the how of creation is anticipated in its newness by the very fluidity of the biblical text by the Bible's use of two distinct statements of human comprehension at the time of writing. As a theoretical deduction from the most careful and massive observation of the creation, the layers and deposits and undulations of this ever-changing old earth, evolution is itself a fluid perception. It raises as many questions as it answers. Evolution represents the best formulation of the knowledge that creation has disclosed to us, but it is the latest word from science, not the last.

If the world is not God's, the most eloquent or belligerent arguments will not make it so. If it is God's world, and this is the first declaration of our creed, then faith has no fear of anything the world itself reveals to the searching eye of science.

Insistence upon dated and partially contradictory statements of how as conditions for true belief in the why of creation cannot qualify either as faithful religion or as intelligent science. Neither evolution over an immensity of time nor the work done in a six-day week are articles of the creeds. It is a symptom of fearful and unsound religion to contend with one another as if they were. Historic creedal Christianity joyfully insists on God as sovereign and frees the human spirit to trust and seek that sovereignty in a world full of surprises.

THE GENERAL CONVENTION
OF THE EPISCOPAL CHURCH

Whereas, the state legislatures of several states have recently passed so-called "balanced treatment" laws requiring the teaching of "Creation-science" whenever evolutionary models are taught; and

Whereas, in many other states political pressures are developing for such "balanced treatment" laws; and

Whereas, the terms "Creationism" and "Creation-science" as understood in these laws do not refer simply to the affirmation that God created the Earth and Heavens and everything in them, but specify certain methods and timing of the creative acts, and impose limits on these acts which are neither scriptural nor accepted by many Christians; and

Whereas, the dogma of "Creationism" and "Creation-science" as understood in the above contexts has been discredited by scientific and theologic studies and rejected in the statements of many church leaders; and

Whereas, "Creationism" and "Creation-science" is not limited to just the origin of life, but intends to monitor public school courses, such as biology, life science, anthropology, sociology, and often also English, physics, chemistry, world history, philosophy, and social studies; therefore be it

Resolved, that the 67th General Convention affirm the glorious ability of God to create in any manner, whether men understand it or not, and in this affirmation reject the limited insight and rigid dogmatism of the "Creationist" movement, and be it further

Resolved, that we affirm our support of the sciences and educators and of the Church and theologians in their search for truth in this Creation that God has given and entrusted to us; and be it further

Resolved, that the Presiding Bishop appoint a Committee to organize Episcopalians and to cooperate with all Episcopalians to encourage actively their state legislators not to be persuaded by arguments and pressures of the "Creationists" into legislating any form of "balanced treatment" laws or any law requiring the teaching of "Creation-science."

67th General Convention of the Episcopal Church, 1982.

Lexington Alliance
of Religious Leaders

The following ministers and religious leaders are very much concerned with and opposed to the possibility of "Scientific Creationism" being taught in the science curriculum of Fayette County Schools.

As religious leaders we share a deep faith in the God who created heaven and earth and all that is in them, and take with utmost seriousness the Biblical witness to this God who is our Creator. However, we find no incompatibility between the God of creation and a theory of evolution which uses universally verifiable data to explain the probable process by which life developed into its present form.

We understand that you may shortly receive considerable pressure from groups advocating the teaching of "Scientific Creationism" alongside of the theory of evolution. However, we feel strongly that to introduce such teaching into our schools would be both divisive and offensive to many members of the religious community of Fayette County, as well as to those not identified with any religious group.

Please be assured of our continuing interest in this issue, and of our strong desire that the Fayette County Public Schools not permit the teaching of "Scientific Creationism" as an alternative "theory" to evolution in science courses.

1981; signed by 78 Kentucky ministers and religious leaders.

THE LUTHERAN WORLD FEDERATION

Symbolic of the prominence of the evolutionary idea in contemporary thought is the occurrence of "evolved" as the last word of the famous closing paragraph of Darwin's *The Origin of Species*, 1859. While not original with the emergence of Darwinism, evolution has nevertheless been intimately associated with it and has in the intervening century become one of the most comprehensive concepts of the modern mind. Consequently the issue cannot be stated in terms of the restricted alternative whether any one phase of evolution (especially the biological) is still "only a scientific theory" or long since "an established fact." Neither is it a matter of holding out the hope that if only enough fault can be found with Darwin the church's doctrine of creation will automatically be accepted and religion can then be at peace with science.

Rather, the evolutionary dynamisms of today's world compel a more realistic confrontation. One area of reality after another has been analyzed and described on the basis of some kind of progressive change until the whole may be viewed as a single process. The standpoint of the one who views this unitary development may be avowedly atheistic in the sense of ruling out the supernatural (Sir Julian Huxley) or just as avowedly Christian in the sense of finding in evolution an infusion of new life into Christianity, with Christianity alone dynamic enough to unify the world with God (Teilhard de Chardin).

In whatever way the process may be ultimately explained, it has come about that an idea which has been most thoroughly explored in the field of biology (lower forms of life evolving into higher) has by means of organismic analogy found universal application. Phenomena thus accounted for range from physical realities (evolution of the atoms and expanding galaxies) to man and his social experience (the evolution of cultural values) including his understanding of time and history (the evolutionary vision of scientific eschatology). Hence there is posited a movement of cumulative change in the organic and the inorganic; in the evolution of life and of man, of social institutions and political constitutions, of emerging races and nations, of language and art forms, of school systems and educational methods, of religion and doctrine; and of science and of the theory of evolution itself.

In the 1959 University of Chicago Centennial Discussions of *Evolution After Darwin* a working definition given to the term evolution was that of a long temporal process, operating everywhere, in which a unidirectional and irreversible natural development generates newness, variety, and "higher levels of organization" (Vol. I, p. 18; Vol. III, p. 111). A noteworthy feature of these discussions was the forthrightness with which at least some of the participants presented evolution in an uncompromising opposition to any notion of the supernatural and in a consistent upholding of naturalistic self-sufficiency in a cosmos which was not created but which has evolved.

With biological evolution (ostensibly a matter of pure science) thereby becoming a metaphysics of evolution it needs to be determined whether religion's proper quarrel is with the science which permits itself such dogmatic extension or whether the misgivings are primarily with the particular philosophical interpretation involved. To the evolutionary concept in general there are however (in spite of innumerable variations) basically two religious reactions.

1. As in the days of the Scopes trial all evolution may still be denied on the grounds of a literalistic interpretation of the Bible, especially Genesis 1-11. Not content with the commitment of faith in the Creator expressed in the First Article of the Apostles' Creed this interpretation may demand a specific answer also to the questions of when creation occurred and how long it took. On the premise of a literal acceptance of the Scriptures as authoritative also in matters of science the whole of past existence is comprehended within the limited time span of biblical chronologies and genealogies. The vastness of astronomical time with its incredible number of light years may be accounted for as an instantaneous arrival of light and the eras of geological and biological time with their strata, fossils, and dinosaurs pointing to the existence of life and death on the earth ages before the arrival of man may be reduced to one literal week of creative activity.

2. On the other hand there are those who can no more close their eyes to the evidence which substantiates some kind of lengthy evolutionary process in the opinion of the vast majority of those scientists most competent to judge than they could deny the awesome reality of God's presence in nature and their own experience of complete dependence upon the creative and sustaining hand of God revealed in the Scriptures. In reference to creation, Langdon Gilkey (*Maker of Heaven and Earth,* 1959, pp. 30 f.) inter-

prets the doctrine as affirming ultimate dependence upon God and distinguishes it from scientific hypotheses which properly deal with finite processes only. Among Lutheran theologians George Forell (*The Protestant Faith*, 1960, p. 109) sees the doctrine of creation not as expressing "a theory about the origin of the world" but as describing man's situation in the world, and Jaroslav Pelikan (*Evolution After Darwin*, Vol. III, p. 31) presents the creation accounts of Genesis as "not chiefly cosmogony" and furthermore sketches a development in the church which by the 19th century had emphasized those aspects of the doctrine of the creation to which Darwin represented a particular challenge and had neglected other important aspects which could be maintained independently of biological research.

An assessment of the prevailing situation makes it clear that evolution's assumptions are as much around us as the air we breathe and no more escapable. At the same time theology's affirmations are being made as responsibly as ever. In this sense both science and religion are here to stay, and the demands of either are great enough to keep most (if not all) from daring to profess competence in both. To preserve their own integrity both science and religion need to remain in a healthful tension of respect toward one another and to engage in a searching debate which no more permits theologians to pose as scientists than it permits scientists to pose as theologians.

Edwin A. Schick, "Evolution", in The Encyclopedia of the Lutheran Church, *Vol. I J. Bodensieck, ed., 1965 Minneapolis: Augsburg Publishing House. The Encyclopedia is a publication of the Lutheran World Federation.*

ROMAN CATHOLIC CHURCH
Pope John Paul II

Cosmogony itself speaks to us of the origins of the universe and its makeup, not in order to provide us with a scientific treatise but in order to state the correct relationship of man with God and with the universe. Sacred Scripture wishes simply to declare that the world was created by God, and in order to teach this truth, it expresses itself in the terms of the cosmology in use at the time of the writer. The sacred book likewise wishes to tell men that the world was not created as the seat of the gods, as was taught by other cosmogonies and cosmologies, but was rather created for the service of man and the glory of God. Any other teaching about the origin and makeup of the universe is alien to the intentions of the Bible, which does not wish to teach how heaven was made but how one goes to heaven.

Address to the Pontifical Academy of Sciences on 3 October 1981

UNITARIAN UNIVERSALIST ASSOCIATION (1977)

Whereas currently there are efforts being made to insert the creation story of Genesis into public school science textbooks; and

*Whereas,*such action would be in direct contradiction with the concept of separation of church and state;

Therefore be it resolved: That the 1977 General Assembly of the Unitarian-Universalist Association goes on record as opposing such efforts.

Be it further resolved: That individual societies are urged to immediately provide petitions on the subject to be signed by members and sent to their legislators; and

Be it further resolved: That this resolution be forwarded to the textbook selection committee of each state department of education by the Department of Ministerial and Congregational Services.

Passed at the 1977 General Assembly of the Unitarian- Universalist Association.

UNITARIAN UNIVERSALIST ASSOCIATION (1982)

W*hereas*, the constitutional principles of religious liberty and the separation of church and state that safeguards liberty, and the ideal of a pluralistic society are under increasing attack in the Congress of the United States, in state legislatures, and in some sectors of the communications media by a combination of sectarian and secular special interests;

Be it resolved: That the 1982 General Assembly of UUA reaffirms its support for these principles and urges the Board of Trustees and President of the Association, member societies, and Unitarian-Universalists in the United States to: ... 2 . Uphold religious neutrality in public education, oppose all government mandated or sponsored prayers, devotional observances, and religious indoctrination in public schools; and oppose efforts to compromise the integrity of public school teaching by the introduction of sectarian religious doctrines, such as "scientific creationism," and by exclusion of educational materials on sectarian grounds. . .

Passed at the 21st annual General Assembly of the UUA in June 1982.
The above excerpt omits other articles of the resolution not directly
related to creationism.

United Church Board for Homeland Ministries: *Creationism, the Church, and the Public School*

I. Background On The Creationism Issue

Creationism is a relatively recent development in an older and on-going controversy concerning the relationship between science and religion. In the 1920's the teaching about evolution in public schools (specifically the work of Charles Darwin) was challenged on the basis of perceived conflict with biblical teaching. In Tennessee John Scopes was convicted of violating a law which made it "illegal … to teach any theory that denies the story of the divine creation of man as taught in the Bible, and to teach instead that man has descended from a lower order of animals." Although the conviction was overturned on a technicality, the Tennessee Supreme Court upheld the constitutionality of the law which was not repealed until 1967.

The central issue in challenges such as this is the apparent conflict between scientific explanations about the origins of life, even the cosmos itself, and biblical accounts of creation. Science and religion often are perceived as being in basic conflict concerning creation.

In more recent decades, the debate has taken a new twist. While still opposing the scientific theories of evolution concerning the origins of life, a number of persons began to suggest that certain scientific data and/or approaches could 'prove' the validity of biblical accounts concerning creation. In the 1960's and early 1970's, several organizations were formed to promote the idea that the creation accounts recorded in the book of Genesis were supported by scientific data. The terms "creation-science," "scientific creationism," and "creationism" are used to describe this interpretation of scripture.

This movement took on more focused activity in 1977 when over twenty state legislatures recorded bills requiring teaching of "creation-science" when evolution was taught. This "balanced treatment" proposition was passed as model legislation by the Arkansas Legislature in 1981.

Opponents of the Act, including religious leaders, educators, and scientists, challenged the constitutionality of the Act in the federal courts (*McLean v Arkansas Board of Education*) and in 1982 the law was declared unconstitutional. A similar law was passed in Louisiana and litigation went all the way to the U.S. Supreme Court. The court in *Edwards v Aguillard* declared the law unconstitutional in 1987. The Supreme Court decision has been applied in subsequent cases involving individual teachers who chose to teach "creation-science" outside the curriculum. Federal courts declared that teaching "creation-science" was a religious advocacy and, therefore, unconstitutional. Courts have taken special care to protect the religious independence of students in the public schools.

Since the Supreme Court decision in *Edwards,* creationists have concentrated their efforts at the level of the local school board, where they pressure educators to teach "creation-science," omit or qualify the teaching of evolution, and/or adopt textbooks that exclude evolution. Additional terms for "creation-science," such as "abrupt appearance theory" or "intelligent design theory" are attempts to avoid the constitutional issue of religious advocacy. However, beyond the notion of "equal time" other issues are emerging. The attempts to use scientific data and methods to prove certain biblical claims are raising concerns among many educators and scientists about the integrity of scientific inquiry itself and what students may be learning about the nature and role of science. Science and scientific methods can be abused by setting out to prove certain assumptions rather than allowing even those assumptions to be open to inquiry and discussion.

The concerns over current activities by creationists touch basic affirmations about the public school made by the United Church Board for Homeland Ministries. The effort to make creationism part of the science curriculum in the public schools tests our commitments to the public school, excellence in education, the integrity of science, and academic freedom. It also tests our interpretation of the Bible and our belief in God's unlimited creative powers.

It is therefore appropriate amidst this controversy for the United Church Board to work with members of the United Church of Christ and others to understand this issue from the perspective of our religious and educational traditions. We mean to assist persons to participate fearlessly in open inquiry, debate, and action concerning the goals of education; to understand the role of science, including an appropriate relationship between science and faith; to help develop consensus in public policy issues affecting the public school; and to support academic freedom at all levels of the educational experience.

II. Affirmations

1. We testify to our belief that the historic Christian doctrine of the Creator God does not depend upon any particular account of the origins of life for its truth and validity. The effort of the creationists to change the book of Genesis into a scientific treatise dangerously obscures what we believe to be the theological purpose of Genesis, viz., to witness to the creation, meaning, and significance of the universe and of human existence under the governance of God. The assumption that the Bible contains scientific data about origins misreads a literature which emerged in a pre-scientific age.

2. We acknowledge modern evolutionary theory as the best present-day scientific explanation of the existence of life on earth; such a conviction is in no way at odds with our belief in a Creator God, or in the revelation and presence of that God in Jesus Christ and the Holy Spirit.

3. We affirm the freedom of conscience and freedom of religion set forth and protected in the U.S. Constitution, including the right of the creationists to their religious beliefs.

4. We believe that the nurturing of faith and religious commitment is the responsibility of the church and home, not of the public school. No person or group should use the school to compel the teaching or acceptance of any creed or to impose conformity to any specific religious belief or practice. Requiring the teaching of the religious beliefs of creationists in the public school violates this basic principle of American democracy. We concur with judicial rulings that the teaching of the religious beliefs of the creationists in the public school science curriculum is unconstitutional.

5. We assert that the public school science curriculum is not the proper arena for the expression of religious doctrine. However, we believe that the public school does have the responsibility to teach about religion, in order to help individuals formulate an intelligent understanding and appreciation of the role of religion in the life and culture of all people and nations. In this context, it is fully appropriate for the public school to include in its non-science curriculum consideration of the variety of religious literature about the creation and origins of human life.

6. We reaffirm our historic commitment to the public school, and declare that each student has the right to an education which rests firmly on the best understandings of the academic community.

7. We affirm our historic commitment to academic freedom in the public school; in that context, the open and full search for truth about all issues in science including creation must proceed in the light of responsible scholarship and research, subject always to the process of peer review, and of factual and logical verification, and of scientific replication.

8. We reject any modification of science textbooks to include the point of view of the creationists or that weakens scientific teachings, and we support publishers who resist this effort. To do otherwise would abridge both academic freedom and the customary practices of careful scholarship.

9. We affirm the responsibility of professional educators to make final decisions about the public school curriculum. These decisions should be based on sound scholarship, competent teaching practices, and policies of local and state school boards which are accountable to the public and in keeping with judicial decisions upholding Constitutional values.

III. RECOMMENDATIONS

1. That through study and discussion we, as church people, become informed about issues of creation raised by both science and religion, including the "creation-science" controversy.

2. That we urge pastors and teachers to preach and teach about issues of creation, particularly the ways of understanding the first eleven chapters of Genesis, the first chapter of the Gospel of John, and other relevant Scripture passages. We further urge pastors and teachers to teach about the problems of biblical literalism in blocking creative dialogue between the faith community and contemporary educational, scientific, and political communities.

3. That we support the determination of schools, school boards, and textbook publishers to retain their professional integrity in treating the creationism issue, carefully recognizing the distinction between promoting religion and teaching about religion.

4. That we make all efforts to resist any viewpoint which would maintain that belief in both a Creator God and in evolutionary theory are in any way incompatible. Confident in our conviction that God is the ultimate source of all wisdom and truth, we encourage the free development of science and all other forms of intellectual inquiry.

5. That clergy and laity exercise their civic responsibility to monitor the work of state legislatures, taking care that any discussion of proposed "creation-science" legislation include educational and constitutional questions, and affirming that such legislation is a violation of the First and Fourteenth Amendments of the U.S. Constitution.

6. That informed persons, including clergy and laity, in each community monitor the work of local school boards and state departments of education, so that issues of 'creation-science" may be discussed fully and openly if and when they come to their agendas. In communities being divided by the creationism controversy, we ask our people to be both a source of reconciliation and a community of support for those who oppose efforts to present creationism as a science.

7. That concerned educators and citizens work with teachers to support their efforts to teach their disciplines with integrity, rather than omit subjects such as evolution as a way of avoiding controversy.

8. That the church renew efforts to understand and relate to science and technology, not only to comprehend and respond to issues of controversy, but also to discover new ways of appreciating and expressing God's creative and redeeming activity.

IV. For Further Reading

Ronald S. Cole Turner, An Unavoidable Challenge: Our Church in an Age of Science and Technology, a Foundation Paper on science and technology as a lifelong issue for education, available from the Division of Education and Publication, UCBHM, Cleveland.

Langdon Gilkey, *Creationism on Trial: Evolution & God at Little Rock*, Harper & Row, 1985.

Betty McCollister, ed., *Voices for Evolution*, the National Center for Science Education, Inc. P.O. Box 9477, Berkeley, CA 94709

October 1992 (This statement supercedes the 1983 statement printed in the first edition of Voices for Evolution)

United Methodist Church

Whereas "scientific" creationism seeks to prove that natural history conforms absolutely to the Genesis account of origins; and,

Whereas, adherence to immutable theories is fundamentally antithetical to the nature of science; and,

Whereas, "scientific" creationism seeks covertly to promote a particular religious dogma; and,

Whereas, the promulgation of religious dogma in public schools is contrary to the First Amendment to the United States Constitution; therefore,

Be it resolved that The Iowa Annual Conference opposes efforts to introduce "scientific" creationism into the science curriculum of the public schools.

Passed June 1984, Iowa Annual Conference of the United Methodist Church.

UNITED PRESBYTERIAN CHURCH IN THE U.S.A. (1982)

EVOLUTION AND CREATIONISM

I. Resolution

Whereas, The Program Agency of the United Presbyterian Church in the USA notes with concern a concerted effort to introduce legislation and other means for the adoption of a public school curriculum variously known as "Creationism" or "Creation Science,"

Whereas, over several years, fundamentalist church leadership, resourced by the Creation Science Research Center and the Institute for Creation Research, has prepared legislation for a number of states calling for "balanced treatment" for "creation-science" and "evolution-science," requiring that wherever one is taught the other must be granted a comparable presentation in the classroom;

Whereas, this issue represents a new situation, there are General Assembly policies on Church and State and Public Education which guide us to assert once again that the state cannot legislate the establishment of religion in the public domain;

Whereas, the dispute is not really over biology or faith, but is essentially about Biblical interpretation, particularly over two irreconcilable viewpoints regarding the characteristics of Biblical literature and the nature of Biblical authority:

Therefore, the Program Agency recommends to the 194th General Assembly (1982) the adoption of the following affirmation:

Affirms that, despite efforts to establish "creationism" or "creation-science" as a valid science, it is teaching based upon a particular religious dogma as agreed by the court (*McLean vs Arkansas Board of Education*);

Affirms that, the imposition of a fundamentalist viewpoint about the interpretation of Biblical literature — where every word is taken with uniform literalness and becomes an absolute authority on all matters, whether moral, religious, political, historical or scientific — is in conflict with the perspective on Biblical interpretation characteristically maintained by Biblical scholars and theological schools in the mainstream of Protestantism, Roman Catholicism and Judaism. Such scholars find that the scientific theory of evolution does not conflict with their interpretation of the origins of life found in Biblical literature.

Affirms that, academic freedom of both teachers and students is being further limited by the impositions of the campaign most notably in the modification of textbooks which limits the teaching about evolution but also by the threats to the professional authority and freedom of teachers to teach and students to learn;

Affirms that, required teaching of such a view constitutes an establishment of religion and a violation of the separation of church and state, as provided in the First Amendment to the Constitution and laws of the United States;

Affirms that, exposure to the Genesis account is best sought through the teaching about religion, history, social studies and literature, provinces other than the discipline of natural science, and

Calls upon Presbyterians, and upon legislators and school board members, to resist all efforts to establish any requirements upon teachers and schools to teach "creationism" or "creation science."

Adopted by General Assembly, 1982.

United Presbyterian Church in the USA (1983)

The Church, the Public School, and Creation Science

Current efforts to legislate the teaching of "creation-science" in the public school challenge and violate basic principles which guide public schools and their responsibility for education of a public that is characterized by its cultural pluralism. These basic principles are grounded both in law (General Welfare Clause of Section 8, Article 1, of U.S. Constitution) and in the Reformed understanding that human response to God's gracious calling is expressed through faithfulness, freedom, and self-determination amidst different claims and alternatives. This Reformed understanding is set forth in the public policy position on public education adopted by the 119th General Assembly:

The biblical impetus toward growth for faith and justice is reaffirmed in the theological stance of the Reformed tradition. This impetus calls for a unique combination of teaching- learning experiences: in home, in church, and in public education.

Persons are called "to glorify God and enjoy him forever." Within the Reformed tradition, this calling is God's act of grace. On the Christian's side the act of grace is affirmed through commitment. But commitment is not simply the acceptance of the truth of certain doctrinal statements. It is much more the embodiment of the lifestyle of Jesus. This embodiment takes place in the everyday struggle to make decisions about the common life of God's creatures. Decision-making implies the freedom of self-determination. It calls for consciousness of alternatives and their consequences. Growth in self-determination is thus best achieved in a setting where alternate loyalties are experienced and reflected upon and where the freedom to create new alternatives is not only permitted but encouraged. Pluralism comprises such a setting, and the public school is the context of pluralism which provides an appropriate atmosphere for growth and development toward the maturity of decision-making and commitment.

In addition, Christian love and respect for persons demand that all persons be free to search for the truth wherever they may find it. This free search for truth which is essential to maturity calls for an appreciation and respect for all human efforts toward justice and love. When public education is not restricted by theological positions or secular ideologies, it provides such an arena for free inquiry and appreciation of all efforts toward humanization.

The Reformed tradition seeks, therefore, to sustain and support all efforts toward the removal of ignorance and bigotry and toward the establishment of free institutions as a source of a high degree of social stability. Public education can be such a free institution where ignorance and bigotry are challenged.[1]

The creation-science controversy thus touches basic tenets that are deeply rooted in the nation and in the Reformed tradition. Our primary intent is to contribute to moral discourse, as these issues are debated within the community of faith as well as within the scientific and educational communities. Our purpose is to help people consider how to think rather than to dictate *what* they are to think.

The goals of this dialogue are to develop public policies which both safeguard individual freedom and contribute to the public good and which strengthen the public school as one of society's most essential institutions, serving all the people. We would mark the discrete functions of the church and the school, while at the same time acknowledging their common commitment to the development of persons and to the formation of a just and humane society.

We accept a responsibility to participate in the education of the public on the issues raised by the creationism controversy and in the continuing formation of public policy affecting the public school. We make these affirmations and offer recommendations for consideration by synods, presbyteries, congregations, and the various publics represented in their membership.

AFFIRMATIONS

1. As citizens of the United States, we are firmly committed to the right and freedom of conscience and freedom of religion, that is, freedom of each citizen in the determination of his or her religious allegiance, and the freedom of religious groups and institutions in the declaration of their beliefs.

2. As Christians, we believe every individual has the right to an education aimed at the full development of the individual's capacities as a human being created by God, including both intellect and character. We also believe that we have the responsibility to educate and thus will seek maximum educational opportunities for

every child of God, that all persons may be prepared for responsible participation in the common life.

3. We affirm that each individual has the right to an education which recognizes rather than obscures the ethnic, racial and religious pluralism of our country and which prepares persons for life in the emerging world culture of the 21st century. Such an education views the individual as a whole person for whom discursive intellect, aesthetic sensitivity and moral perspective are intimately related.

4. We re-affirm our historic commitment to the public school as one of the basic educational institutions of the society. We celebrate its inclusiveness and its role as a major cohesive force, carrying our hopes for a fully democratic and pluralistic society. We further re-affirm the responsibility of public institutions to serve all the population as equitably as possible, neglecting none as expendable or undeserving of educational opportunity.

5. We affirm our faith that God is the author of truth and the Holy Spirit is present in all of our common life, to lead us all into truth. Ours is a journey of faith and of revelation in which the human spirit is fed and led but not coerced.

6. We believe that the nurturing of faith is the responsibility of the home and the church, not the public school. Neither the church nor the state should use the public school to compel acceptance of any creed or conformity to any specific religious belief or practice.

7. We affirm the professional responsibility of educators to make judgments about school curriculum which are based on sound scholarship and sound teaching practices.

8. We affirm that it is inappropriate for the state to mandate the teaching of the specific religious beliefs of the creationists in accord with the Overton ruling (*McLean vs Arkansas Board of Education*). We also affirm the responsibility of the public school to teach about religious beliefs, ideas and values as an integral part of our cultural heritage. We believe the public school has an obligation to help individuals formulate an intelligent understanding and appreciation of the role of religion in the life of people of all cultures. In the context of teaching about religion, it is appropriate to include in the public school curriculum consideration of the variety of religious interpretations of creation and the origins of human life.

9. We affirm our uncompromising commitment to academic free-dom, that is, freedom to teach and to learn. Access to ideas and opportunities to consider the broad range of questions and experiences which constitute the proper preparation for a life of responsible citizenship must never be defined by the interests of any single viewpoint or segment of the public.

10. We acknowledge the need to enlarge the public participation in open inquiry, debate and action concerning the goals of education, and in the development of those educational reforms which equip children, youth and adults with equal opportunities to participate fully in the society. This participation must respect the constitutional and intellectual rights guaranteed school personnel and students by our law and tradition.

11. We pledge our continuing efforts to strengthen the public school as the most valuable, open, and accessible institution for formal education for all the people; we assert that educational needs are more important than economic, political and religious ideologies as the basis upon which to formulate educational policies.

12. We affirm anew our faith and oneness in Christ, the way, the truth and the life, as we struggle to make a faithful witness amid the conflict of convictions and conclusions between sisters and brothers who bear a common name.

RECOMMENDATIONS

For Congregations

1. That the General Assembly encourage congregations to study the issues in the creation-science controversy, giving particular attention to:

   the historic role of the churches in the founding and developing of the public school.

   the diversity of belief about creation and human origin present in our society.

   the principles and assumptions which guide the development of the science curriculum in the public school and the use of scientific inquiry within all disciplines and subjects.

   the essentials of the church-state issues as they apply to the public school, including a review of the major U.S. Supreme Court decisions and the recent court decisions on the creationism issue (i.e. *McLean vs Arkansas Board of Education*).

the processes of policy-making for the public school including the appropriate roles of the community, the educator, the parent, and the church.

2. That the General Assembly urge congregations to encourage local school boards to discuss issues of creation-science fully and openly, if and when they come onto the board's agenda.

3. That the General Assembly urge congregations to encourage and assist teachers and administrators in becoming sensitive to the religious perspectives of all persons in the schools, without sacrificing their professional commitments and standards regarding the teaching of science and teaching about religion.

4. That the General Assembly encourage congregations in communities divided by the creationism controversy to work for reconciliation and to provide a community of support for those struggling to keep the schools free of ideological indoctrination.

5. That the General Assembly encourage pastors and Christian educators to help their congregations to interpret the biblical passages dealing with creation and the origins of human life in ways that take their message seriously.

6. That the Mission Board provide study resources including the study paper prepared by the United Ministries in Education, "Creationism, the Church, and the Public School." (The paper is available from United Ministries in Education, c/o American Baptist Churches, Valley Forge, PA 19481.)

7. That the General Assembly commend the paper, "The Dialogue Between Theology and Science" (adopted by the 122nd General Assembly), as a study document addressing the basic issues related to the ongoing debate regarding the teaching of evolution and creationism in public schools.

For Synods and Presbyteries

8. That the General Assembly encourage synods and presbyteries to give attention to the work of state legislatures and their committees, taking care that any discussion of proposed creation-science legislation include broader educational, religious, and constitutional questions, and to join with others to have creation-science legislation declared unconstitutional when it is

in violation of the First and Fourteenth Amendments to the U.S. Constitution.

9. That the General Assembly urge synods and presbyteries to encourage educators and citizens to examine the textbooks being used now in the public schools for the adequacy of their teaching about creation and evolution and about the differing religious perspective and interpretations of origins, and to resist every effort to purge or discredit data which are held to be part of our common history and heritage.

10. That the General Assembly encourage presbyteries to provide in resource centers information about creation-science, evolution-science and related public school issues.

Footnote in the original:
[1] Minutes of the 119th General Assembly, p. 526. The paper was adopted by the General Assembly and commended to the Church for study.
Passed at the 195th General Assembly of the United Presbyterian Church in the U.S.A., 1983.

PART FOUR

Educational Organizations

AMERICAN ASSOCIATION
OF PHYSICS TEACHERS

The Council of the American Association of Physics Teachers opposes proposals to require "equal time" for presentation in public school science classes of the religious accounts of creation and the scientific theory of evolution. The issues raised by such proposals, while mainly focusing on evolution, have important implications for the entire spectrum of scientific inquiry, including geology, physics, and astronomy. In contrast to "Creationism," the systematic application of scientific principles has led to a current picture of life, of the nature of our planet, and of the universe which, while incomplete, is constantly being tested and refined by observation and analysis. This ability to construct critical experiments which can result in the rejection or modification of a theory is fundamental to the scientific method. While our association does not support the teaching of oversimplified or dogmatic descriptions of science, we also reject attempts to interfere with the teaching of properly developed scientific principles or to introduce into the science classroom religious or mystical concepts that have no logical connection with observed facts or with widely accepted scientific theories. We therefore strongly oppose any requirement for parallel treatment of scientific and non-scientific discussions in science classes. Scientific inquiry and religious beliefs are two distinct elements of the human experience. Attempts to present them in the same context can only lead to misunderstandings of both.

Approved by the Council of the American Association of Physics Teachers on 26 January 1982. Identical to the text of the statement of 15 December 1981 by the American Physical Society.

AMERICAN ASSOCIATION OF UNIVERSITY WOMEN

The American Association of University Women is committed to the pursuit of knowledge and access to that knowledge by all citizens. AAUW is also committed as a national organization to the doctrine of separation of church and state. We are concerned that the inclusion in the public schools of information on the creationist theory will open the door to rightful requests for equal time by the many individual faiths, thus creating an unmanageable situation. Decisions need to be made relating to questions such as:

Who is qualified to relay this information to students?
Who will decide what texts to recommend for further reading?
Which theories will be included for presentation?

AAUW recognizes that theory will not be taught in the classroom, but we have reservations as to how it will be presented. Is it not better to leave the responsibility of religious thought to individual churches? All knowledge is not gained in the public classroom. AAUW believes citizens have a protected right to avail themselves of education through many sources, and the primary source for religious education must be the church.

ASSOCIATION OF PENNSYLVANIA STATE COLLEGE AND UNIVERSITY BIOLOGISTS

Throughout the United States, "Scientific Creationism," a religious doctrine based upon the literal interpretation of the Bible, is being proposed as a valid scientific alternative to the Theory of Evolution. Creationists who represent this fundamentalist Christian religious movement are seeking "equal time" in science classrooms and science textbooks.

The Creationists' movement is an attempt to persuade, mislead, and pressure legislators, public school officials and the general public that since evolution is "only" a theory, implying opinion or conjecture, it is therefore open to any alternative. They propose that their alternative, the "Theory of Special Creation," is scientific and therefore is just as valid as the Theory of Evolution. Creationists reject the evolution of life from a single line of ancestors through chance mutation and natural selection and hold that the universe and all living things were divinely created beginning six to ten thousand years ago. They cite as their "scientific evidence" the biblical story of Genesis as written in the King James version of the Bible. Although Creationists are attempting to equate "Special Creation" as a scientific theory, they in fact claim absolute truth for their belief. Science, which does not deal with beliefs based on faith and does not claim absolute truth for its findings, utilizes an organized method of problem solving in an attempt to explain phenomena of our universe.

The Association of Pennsylvania State College and University Biologists together with other scientific associations such as the National Association of Biology Teachers, the National Academy of Science, the American Association for the Advancement of Science and the American Institute for Biological Sciences agrees that "Scientific Creationism" does not meet the criteria of science and cannot be considered a scientific theory. Scientists of these associations agree that Creationism can be neither verified nor refuted through scientific investigation, and the models or beliefs which involve the supernatural are not within the domain of science. However, to support the Theory of Evolution is not to be "antireligious" as Creationists propose. The majority of religions in America find no basic conflict between religion and science, and most accept the Theory of Evolution and reject Creationism. Throughout the U.S. scientists as well as clergy have opposed the Creationists' attempt to legislate the teaching of "Scientific Creationism" in science classrooms. During the December 1981 trial in Arkansas, in which a Creationist "equal time" law was contested and overturned, a great majority of witnesses in support of the Theory of Evolution were clergy of the Catholic, Protestant, and Jewish faiths.

The Theory of Evolution meets the criteria of science and the criteria of a scientific theory and is not based on faith, mere speculation or dogma. Evolution as a scientific theory is supported by a vast body of scientifically scrutinizable evidence coming from such sources as anatomy and physiology, biochemistry, genetics and the fossil record. To state, as Creationists do, that the Theory of Evolution is "only" a theory illustrates ignorance of science and the scientific method. The Theory of Evolution will be accepted and supported by the scientific community unless another theory which is based on science and the scientific method takes its place.

The Association of Pennsylvania State College and University Biologists recognizes that the move to equate a non- scientific belief with science is a threat to the very integrity of science. APSCUB respects the religious beliefs held by Creationists and others pertaining to the origin and diversity of life and does not oppose the teaching of those concepts as religion or philosophy. However, APSCUB members as scientists and educators are in opposition to any attempt to introduce Creationism or any other non-scientific or pseudoscientific belief as science in the public school system in the Commonwealth of Pennsylvania. APSCUB further recommends the following:

1. All public school science teachers in the Commonwealth should reject science textbooks which treat Creationism as science. The inclusion of non-scientific material as science in a science textbook reflects on the credibility of the teacher who uses it. Textbooks which deal with the diversity of life but do not mention the Theory of Evolution or restrict its discussion should also be rejected.

2. Biology teachers in the public school system of Pennsylvania should teach the Theory of Evolution not as absolute truth but as the most widely accepted scientific theory on the diversity of life. Biology teachers of the Commonwealth should not be intimidated by pressures of the Creationists and simply avoid the issue by not teaching the Theory of Evolution. Avoiding established concepts in science is pseudoscience which also threatens the integrity and credibility of science. Avoiding the teaching of evolution is a victory for the Creationists.

Members of APSCUB will, when possible, give advice and support to teachers, legislators, public school officials, and the general public where matters of "Scientific Creationism" or other non-scientific beliefs concerning the diversity of life arise in their local community within the Commonwealth of Pennsylvania.

Undated; 1982 or later.

AUBURN UNIVERSITY
FACULTY SENATE (1981)

W e understand that the Alabama legislature is considering a requirement that "Scientific Creationism" be included as an alternative to evolutionary theory during discussions in Alabama public schools of the origin and development of life; and

We consider the theory of scientific creationism to be neither scientifically based nor capable of performing the roles required of a scientific theory; and

We agree with the statement of the National Academy of Sciences that "religion and science are separate and mutually exclusive realms of human thought whose presentation in the same context leads to misunderstanding of both scientific theory and religious belief"; and

The proposed action would impair the proper segregation of teaching of science and religion to the detriment of both; and

We favor the continued observance of the First Amendment to the U.S. Constitution guaranteeing freedom of religion by assuming separation of Church and State; and

The inclusion of the theory of creation represents dictation by a lay body of what shall be included within science;

Therefore, let it be resolved that the Auburn University Senate go on record in strenuous opposition to any legislative attempt to determine or to direct what is taught as science in Alabama's public schools.

A variation of the University of Alabama, Huntsville, faculty senate resolution adapted and ratified by voice vote, without dissent, by the Auburn University faculty senate on 10 March 1981. Wording is inferred from the Hunstville resolution and a memorandum attached to it from John Kuykendall to Delos McKown spelling out the changes made at Auburn.

AUBURN UNIVERSITY
FACULTY SENATE (1983)

To: Members of the Science Work Group who developed the 1982 revision of the Science Course of Study

We, the undersigned members of the Auburn University faculty in the sciences, are writing to express our dismay at the action of your committee in removing references to standard topics and concepts in the fields of biological and earth sciences from the Alabama Course of Study of Science.

Recent reports from study groups have emphasized the great deficiencies in science education across the nation. We who teach the graduates of Alabama high schools are particularly aware that our state is no exception. Lawmakers and civic and business leaders alike agree that Alabama must develop "high-tech" industries if we are to prosper or even keep up with our neighbors economically. Yet we are seeing the undermining of teaching of science in the public school to such an extent that few of our best and brightest students are likely to be directed toward careers in science and engineering. Those who are will enter college woefully unprepared to think scientifically and lacking the basic acquaintance with current ideas and facts in science on which a college teacher expects to build.

The signers of this letter represent a wide spectrum of religious beliefs as well as a wide variety of scientific disciplines. Our concern is not with the beliefs of individuals, but with what is genuine science, and that Alabama students be exposed to the scientific information and ideas on which the modern technological world is based. The Course of Study as currently stated gives so much leeway that a course called "biology" or "earth science" could be taught with no scientific content at all. We must not handicap Alabama students with that possibility!

We do not know how you voted on the question of removing terms relating to evolution, the history of the earth, and the age of the universe from the Course of Study. We do know that standard parliamentary procedure allows one who voted for a motion to move for its reconsideration. We urge you to take this or whatever other means lie at your disposal to reconsider the damaging position previously taken — for the sake of Alabama young people and the welfare of our State as a whole.

Passed by the University Senate.

BIOLOGICAL SCIENCES CURRICULUM STUDY (1971)
The BSCS Position on the Teaching of Biology

D r. Addison E. Lee, Professor of Science Education and Biology, and Director of the Science Education Center, The University of Texas at Austin, serves as Chairman of the Board of Directors of the Biological Sciences Curriculum Study. His distinguished accomplishments as science educator and biologist enable him to write with authority in support of the BSCS position on the teaching of evolution. Dr. Lee's many publications as author or editor include Laboratory Studies in Biology and a monograph series entitled Research and Curriculum Development in Science Education.

The BSCS program began in 1959 amid considerable debate about the approach to be taken in the teaching of biology. Should it be molecular, organismal, developmental, ecological, or other? Should it include one textbook or several? How much and what kind of attention to laboratory work should be given? Amidst all these debates, however, it was an early consensus that certain themes should be included in all biology programs, no matter what approach is selected, and whatever attention may be given to various details. These themes were identified and have consistently pervaded the several approaches and different materials developed by the BSCS during the past twelve years. They are:

1. Change of living things through time: evolution

2. Diversity of type and unity of pattern in living things

3. The genetic continuity of life

4. The complementarity of organism and environment

5. The biological roots of behavior

6. The complementarity of structure and function

7. Regulation and homeostasis: preservation of life in the face of change

8. Science as inquiry

9. The history of biological conceptions

It should be noted that these unifying themes were identified and accepted by a large group of distinguished scientists, science teachers, and other educators. And although members of this group represented many interests, specialties, and points of view, there was and has continued to be general agreement concerning the importance, use, and nature of these themes.

It should also be noted that evolution is not only one of the major themes but is, in fact, central among the other themes; they are inter-related, and each is particularly related to evolution.

The position of the BSCS on the importance of evolution in teaching biology has been clearly stated in both the first (1963) and second (1970) editions of the Biology Teachers' Handbook:

It is no longer possible to give a complete or even a coherent account of living things without the story of evolution. On the other hand, many of the most striking characteristics of living things are "products" of the evolutionary process. We can make good sense and order of the similarities and differences among living things only by reference to their evolution. The relations of living things to the particular environments in which they live, their distribution over the surface of the earth, the comings and goings of their parts during development, even the chemistry by which they obtain energy and exchange it among their parts — all such matters find illumination and explanation, in whole or in part, from the history of life on earth.

On the other hand, another great group of characteristics of living things can be fully understood only as the means and mechanisms by which evolution takes place. There are first, and conspicuously, the events of meiosis and fertilization, universal in sexual reproduction. It is only in terms of the contribution of these processes to the enhancement and sorting out of a vast store of heritable variations that we make sense of them. The same point applies to the complex processes that go under the name of mutation. Similarly, we see everywhere the action and consequences of natural selection, of reproductive isolation of populations, of the effects of size and change on intrabreeding groups.

Evolution, then, forms the warp and woof of modern biology...[1]

Evolution is a scientific theory in the sense that it is based on scientific data accumulated over many years and organized into a unifying idea widely accepted by modern biologists. The BSCS is concerned with any scientific theory relevant to the biological sciences that can be dealt with in terms of scientific data accumulated and organized. It is not, on the other hand, concerned with religious doctrines that are based only on faith or beliefs, nor does it consider them relevant to the teaching of biological science.

The BSCS program was carried through an extensive tryout period during its early development; feedback and input from hundreds of scientists and science teachers were used in the initial edition that was made available to biology teachers in the United States. A revised second edition of the three major textbooks produced has been published, and a revised third edition is nearing completion. In spite of efforts of various groups to force changes in the content of the texts by exerting pressures on textbook selection committees and on local and state governments, throughout the last twelve years the BSCS position on using the unifying themes of biology remains unchanged.

Footnote in original:
[1] *BSCS, Biology Teachers' Handbook, Joseph J. Schwab (supervisor), John Wiley and Sons, New York, 1963. BSCS, Biology Teachers' Handbook, Second Edition, Evelyn Klinckmann (supervisor), John Wiley and Sons, New York, 1970.*

BIOLOGICAL SCIENCES CURRICULUM STUDY (1995) Position on the Teaching of Evolution for Voices for Evolution

B SCS, founded in 1958, was largely responsible for reintro-
ducing evolution into the high school biology curriculum,
following a four-decade period during which evolution virtually disappeared from high school biology textbooks. From its
inception, BSCS has treated evolution as the central organizing
theme of biology, listing it first, for example, among the biological
principles that guided the development of all early BSCS programs.

The Biology Teachers' Handbook, published by BSCS in 1963,
stressed the importance of concentrating on major principles in biology and gave special attention to evolution, stating: " It is no longer
possible to give a complete or even a coherent account of living
things without the story of evolution." The intervening three decades
have affirmed that assertion, with progress in genetics, molecular biology, behavior, development, neuroscience, and other sub-
disciplines reinforcing and expanding evolutionary perspectives
originally based on gross morphological data.

The recent and rapid growth of knowledge in all areas of biology
makes it ever more important — and difficult — to focus curriculum
and teaching on major principles. To that end, BSCS recently published Developing Biological Literacy: A Guide to Developing Secondary and Post-secondary Biology Curricula (1993). This
document identifies six unifying principles of biology that should
pervade the teaching of biology, and it states the BSCS position on
evolution quite clearly:

> How can one simultaneously account for the extraordinary
> diversity and observable unity of living systems in the world
> today? The answer, in a word, is evolution. Evolution is the
> unifying theory of biology because it has played a role in the
> history and lives of all living organisms on Earth today — and
> of those that are now extinct. Evolution is the major concep-
> tual scheme of biology because it helps us understand relation-
> ships between organisms, past and present, and the many ways
> organisms have succeeded in different habitats.

We recognize that there are other ways of knowing, but ours is the scientific pursuit of knowledge. As BSCS approaches its fortieth anniversary of service to science education, it remains committed to the accurate and thorough representation of evolution as the conceptual keystone to our understanding of life on Earth. Furthermore, BSCS will continue to defend scientific integrity and will resist all attempts to influence its materials in ways that portray non-scientific explanations of life on Earth as scientifically valid.

Approved by the BSCS Board of Directors
January 1995

COMMITTEE FOR SCIENTIFIC INVESTIGATION OF CLAIMS OF THE PARANORMAL (CSICOP)
Position Statement on
The Teaching of Evolution and the "Scientific" Creationist Challenge

Evolution is the organizing principle of modern biology and is as well established in science as are the principle of gravity and the fact that the earth orbits the sun. Contemporary scientists around the world agree, whatever their national, religious, or cultural affiliations.

Although scientists disagree about such things as the rates, dates, and mechanisms of evolution, virtually no active scientist challenges the fact that evolution has occurred. Furthermore, the fact that scientists debate aspects of evolution is a strong sign that evolution is a healthy science that has not lain dormant in the century since Darwin's death.

The anti-evolutionist "scientific" creationists promote a social and political movement, not a scientific one. They are attempting to impose a sectarian religious view, the literal interpretation of Genesis, upon the public schools. But this is not all: they are claiming to be able to scientifically demonstrate that the world was created suddenly, all at once, a relatively short time ago. Competent scientists, including many Fellows and Scientific Consultants of CSICOP, who have examined the claims of "scientific" creationism, have found them baseless. There is no scientific evidence supporting the instantaneous creation of the earth and all the creatures on it but there is much evidence from many scientific fields that the universe has changed extensively through time.

The gains of the "scientific" creationists have been made through political pressure rather than through the scientific acceptability of their ideas. Science and science literacy suffer greatly when science is subordinated to political pressure. For scientific literacy to increase among Americans, science rather than pseudoscience must be taught to children.

CSICOP urges the public and the mass media to recognize "scientific" creationism as a narrow, religiously based lobby, not a science, and to seek out expert opinion outside the creationist camp when confronted by creationist pseudoscientific claims.

1994

GEORGIA CITIZENS'
EDUCATIONAL COALITION

STATEMENT ON THE TEACHING OF CREATIONISM
IN GEORGIA PUBLIC HIGH SCHOOL SCIENCE CLASSES

There is grandeur in this view of life, with its several powers, having been originally breathed by the Creator into a few forms or into one ... and from so simple a beginning endless forms most beautiful and most wonderful have been, and are being evolved.

— Charles Darwin
The Origin of Species

We oppose the teaching of "creationism" as science in Georgia's public schools.

Creationism is based on the religious belief in biblical literalism, or biblical inerrancy, and not on scientific theory. It includes belief in six 24-hour days of creation which occurred less than 10,000 years ago.

The First Amendment specifically forbids the State to force its citizens to profess a belief, or disbelief, in any religion. Creationism is a particular sectarian doctrine held only by those who believe in biblical literalism.

We have no objection to the belief in biblical literalism by those who are obliged by their religion to do so, but object strongly to injecting this religious belief, in the form of creationism, into the science classroom.

However, we recognize the right of parents to uphold their deep religious convictions by withdrawing their children from the study of the scientific theory of evolution.

Many of us believe there is no contradiction between the acts of the Creator God in the Bible and the theory of evolution, and in fact see the evolutionary process as one of God's greatest works.

It is no longer possible to teach biology without the study of the scientific theory of evolution, which has been universally accepted into mankind's general body of knowledge, and stands today as the organizing principle of biology and the general theory of life. There is no competing theory that is taken seriously.

We therefore strongly oppose the teaching of creationism in Georgia's public high school science classrooms because

1. it is not science, and

2. it would impose a particular religious belief on our students.

1980; written by Charles C. Brooks, President.

IOWA COUNCIL OF SCIENCE SUPERVISORS

Because of the insistence that special creation be taught in Iowa science courses as an alternative concept to evolution, we, the Iowa Council of Science Supervisors, as representatives of the science educators in Iowa, make the following statement:

Science educators are responsible for interpreting the spirit and substance of science to their students. Teachers are bound to promote a scientific rationale based upon carefully defined and objective judgments of scientific endeavors. When conflicts arise between competing paradigms in science, they must be resolved by the scientific community rather than by the educators of science.

Based upon court decisions in Indiana and Tennessee, and in the creationists' own statements of beliefs, the Creation Research Society is premised upon the full belief in the Biblical record of special creation.

"The Bible is the Written Word of God, and because it is inspired throughout, all its assertions are historically and scientifically true in all original autographs. To the student of nature this means that the account of origins in Genesis is a factual presentation of simple historical truths." [1]

Science is tentative and denies an ultimate or perfect truth as claimed by scientific creationism. We suggest that creationists submit their creation theories and models to recognized science organizations such as the American Association for the Advancement of Science (AAAS) or their affiliated scientific societies. The claims of these paradigms should be substantiated with validated objective evidence. The scientific organizations would assume responsibility for analyzing the materials, making their findings available for national review through AAAS scientific journals.

Until "scientific creation" receives substantial support from such organizations as AAAS, American Anthropological Association, state academies of science, National Academy of Science, and national paleontological and geological associations, it is recommended that this organization and the science teachers of Iowa reject further consideration of scientific creationism as an alternative approach to established science teaching practices.

---

[1] *Membership application forms for the Creation Research Society, Wilbert H. Rusch, Membership Secretary, 2712 Cranbrook Road, Ann Arbor, Michigan 48104. Corrections of spelling and punctuation by editors.*

Iowa Department of Public Instruction

Creation, Evolution and Public Education: The Position of the Iowa Department of Public Instruction

The Controversy

In Iowa and other states, "creationism" has recently been advanced as an alternative to the theory of evolution. Attempts have been made to legislatively mandate "equal time" for creationist concepts in science classrooms, materials, and textbooks.

Interviews and surveys conducted by the Iowa Department of Public Instruction show that most Iowa religious leaders, science educators, scientists and philosophers contacted support the present patterns of teaching science in Iowa's schools. In addition, due to the nature of scientific and theological concepts, these authorities feel that the specifics of each discipline should be confined to their respective houses.

The National Academy of Science has stated that religion and science are "separate and mutually exclusive realms of human thought whose presentation in the same context leads to misunderstanding of both scientific theories and religious beliefs."[1]

Creationism

In America, religion is usually defined as the expression of man's belief in, and reverence for, a metaphysical power governing all activities of the universe. Where there is not belief in metaphysical power, religion is a concern for that which is ultimate. Generally, creationism is a religious concept. It proposes that all living things were created by a Creator. According to the creation model, "all living things originated from basic kinds of life, each of which was separately created."[2]

There are many versions of creation. Generally, creationists advocate that all permanent, basic life forms originated thousands of years ago through directive acts of a Creator — independent of the natural universe. Plants and animals were created separately with their full genetic potentiality provided by the Creator. Any variation, or speciation, which has occurred since creation has been within the original prescribed boundaries. Since each species contains its full potentiality, nature is viewed as static, reliable and predictable. Based on alleged gaps in the geologic record, creationists reject the theory of the descent of plants and animals from a single line of ancestors arising through random mutation and successively evolving over billions of years. It is further alleged that, through analysis of geologic strata, the earth has experienced at least one great flood or other natural global disaster accounting for the mass extinction of many biological organisms. Following such extinctions there followed sudden increases in the number, variety and complexity of organisms.

Having all Biblical accounts of creationism placed in comparative theology courses with other religious accounts of origins will not placate ardent creationists. They require that creationism be presented as a viable scientific alternative to evolution.[3] More zealous creationists argue that "it is only in the Bible that we can possibly obtain any information about the methods of creation, the order of creation, the duration of creation, or any other details of creation."[4]

SCIENCE

Science is an attempt to help explain the world of which we are a part. It is both an investigatory process and a body of knowledge readily subjected to investigation and verification. By a generally accepted definition, science is not an indoctrination process, but rather an objective method for problem solving. Science is an important part of the foundation upon which rest our technology, our agriculture, our economy, our intellectual life, our national defense, and our ventures into space.

The formulation of theories is a basic part of scientific method. Theories are generalizations, based on substantial evidence, which explain many diverse phenomena. A theory is always tentative. It is subject to test through the uncovering of new data, through new experiments, through repetition and refinements of old experiments, or through new interpretations. Should a significant body of contrary evidence appear, the theory is either revised or it is replaced by a new and better theory. The strength of a scientific theory lies in the fact that it is the most logical explanation of known facts, principles, and concepts dealing with an idea which does not currently have a conclusive test.

EVOLUTION

The theory of evolution meets the criteria of a scientific theory. It can explain much of the past and help predict many future scientific phenomena. Basically, the theory states that modern biologic organisms descended, with modification, from pre-existing forms which in turn had ancestors. Those organisms best adapted, through anatomical and physiological modification to their environment, left more offspring than did non-adapted organisms. The increased diversity of organisms enhanced their ability to survive in various environments and enabled them to leave more progeny.

The theory of evolution is designed to answer the "how" questions of science and biological development; it cannot deal effectively with the "who" or "why" of man's origin and development. It is, however, an effective means of integrating and clarifying many otherwise isolated scientific facts, principles and concepts.

There have been alternatives proposed to the theory of evolution (i.e., creationism, exo-biology, spontaneous generation); however, none are supported by the amount of scientific evidence that presently supports the theory of evolution.

It is evident that the *process* of evolution occurs. Successful species of living organisms change with time when exposed to environmental pressures. Such changes in species have been documented in the past, and it can be confidently predicted that they will continue to change in the future. Evolution helps explain many other scientific phenomena: variations in disease, drug resistance in microbes, anatomical anomalies which appear in surgery, and successful methods for breeding better crops and farm animals. Modern biological science and its applications on the farm, in medicine, and elsewhere are not completely understandable without many of the basic concepts of evolution.

There are many things that evolution is not. It is not dogma. Although there is intense dispute among scientists concerning the details of evolution, most scientists accept its validity on the ground of its strong supporting evidence.

DEPARTMENT OF PUBLIC INSTRUCTION DECISION

Teaching religious doctrine is not the science teacher's responsibility. Teachers should recognize the personal validity of alternative beliefs, but should then direct student inquiries to the appropriate institution for counseling and/or further explanation. Giving equal emphasis in science classes to non-scientific theories that are presented as alternatives to evolution would be in direct opposition to understanding the nature and purpose of science.

Each group is fully entitled to its point of view with respect to

the Bible and evolution; but the American doctrine of religious freedom and the Establishment Clause in the First Amendment to the
U.S. Constitution forbid either group — or any other religious group
— from pressing its point of view on the public schools. An Indiana
court decision declared: "The prospect of biology teachers and students alike forced to answer and respond to continued demand for
'correct' Fundamentalist Christian doctrines has no place in public
schools." [5]

The science curriculum should emphasize the theory of evolution as a well-supported scientific theory — not a fact — that is taught
as such by certified science teachers. Students should be advised that
it is their responsibility, as informed citizens, to have creationism
explained to them by theological experts. They must then decide for
themselves the merits of each discipline and its relevance to their
lives.

The Iowa Department of Public Instruction feels that public schools cannot be
surrogate family, church and all other necessary social institutions for students,
and for them to attempt to do so would be a great disservice to citizens and
appropriate institutions.

Footnotes in original:
[1] Resolution adopted by the National Academy of Science
and the Commission of Science Education of the American Academy [sic]
for the Advancement of Science (Washington, D.C. 17 October 1972).
[2] Bliss, R. B., Origins: Two Models: Evolution, Creation (San Diego:
Creation Life Publishers, 1976), p. 31.
[3] Morris, Henry M., The Remarkable Birth of Planet Earth (San Diego:
Creation Life Publishers, 1972).
[4] National Association of Biology Teachers, A Compendium of Information
on the Theory of Evolution and the Evolution-Creationism Controversy
(June 1977).
[5] Hendren vs Campbell, Supreme [sic] Court No. 5, Marion County,
Indiana (1977), p. 20
Released by the Iowa DPI in March 1980.

MICHIGAN SCIENCE TEACHERS ASSOCIATION
Creation, Evolution, and Science Education

Scientific creation, special creation, and creation-science are terms used synonymously when referring to the thesis that the universe and all forms of life were brought into existence by sudden acts of a Divine Creator. Supporters of this thesis are creationists, some of whom are campaigning vigorously in favor of the inclusion of creation-science in the science classrooms of the nation's public schools. In effect, such inclusion would constitute a two-model approach to questions of origins. One of these models is the *theory of evolution*; the second model is *creation-science*.

The theory of evolution is the theory or model presented in the life sciences curricula of public school science classrooms. Evolution theory is taught because its existence as a developing network of observations, hypotheses, predictions, facts, principles, and sub-theories is the result of scientific inquiry free of any a priori design.

In comparison, the creation-science model is not an observation-hypothesis-prediction-fact-principle-accessory theory sequence. It does not encourage open-ended questioning because any raising of questions must produce answers that converge on the Divine Creator thesis. As a result, the creation-science model cannot generate information and ideas useful in the development of new areas of scientific investigation. Thus, creation-science is not acceptable as a scientific theory. Even so, the necessity exists for public school science educators to consider certain causal concerns of creationists. A major concern is that students from creationist backgrounds are exposed to theories regarding questions of origins not consonant with their religious training; a second concern is that many science teachers teach the theory of evolution as fact.

Therefore, in consequence of the creationists' concerns and in consequence of the need to maintain the integrity of science education, the Michigan Science Teachers Association adopts the following position with respect to the evolution/creation issue:

1. The Michigan Science Teachers Association affirms the necessity of rejecting the teaching of non-scientific theories in the science classrooms of Michigan's public schools.
2. The Michigan Science Teachers Association recommends that its professional development committee be responsible for the design of an inservice model for helping science teachers learn how to work sensitively and objectively with the evolution/creation concerns expressed by students, parents, and boards of education.

3. The Michigan Science Teachers Association reaffirms its goals of a) helping students acquire useful science knowledge and skills b) helping students progress in the understanding and use of processes of scientific inquiry, and c) helping students separate scientific thought and activity from thought rightfully the province of humankind's diverse ways of spiritual expression and responsiveness to the need for authoritarian guidance.

4. The Michigan Science Teachers Association recommends the establishment of procedures for the dissemination of the position expressed herein to Michigan Science teachers, to the Michigan Department of Education through its science specialist, and to Michigan Boards of Education and to science specialists of other states on request.

Approved on November 21, 1981, by the Board of Directors of the Michigan Science Teachers Association on behalf of the Michigan Science Teachers Association.

NATIONAL ASSOCIATION OF
BIOLOGY TEACHERS: *Scientific Integrity*

The ongoing procedures and processes of science are well defined within each scientific discipline, including biology. The principles and theories of science have been established through repeated experimentation and observation and have been refereed through peer review before general acceptance by the scientific community. Acceptance does not imply rigidity or constraint, or denote dogma. Instead, as new data become available, scientific explanations are revised and improved, or rejected and replaced. Materials, methods, and explanations that fail to meet these ongoing tests of science are not legitimate components of the discipline and must not be part of a science curriculum.

Science may appear to conflict with other ways of knowing about the universe, unfortunately leading some groups to see selected theories of science as a threat to their belief systems. This is not the case; science does not, in fact cannot, study, explain, or judge non-scientific issues or supernatural belief systems.

Science is but one way of making sense of the world, with internally-consistent methods and principles that are well described. Among these principles is the notion that proposed causes and explanations must be naturalistic. Any attempt to mix or contrast supernatural beliefs and naturalistic theories within science misrepresents the scientific enterprise and debases other, non-scientific, ways of knowing. These attempts, which commonly result from a misunderstanding of the nature of science itself, have no place in science or in the science classroom or laboratory.

The credibility and utility of science, and therefore biology, depend on maintaining its integrity. NABT has a special obligation to promote this integrity in life science education. The data, concepts, and theories of science presented to students must meet the accepted standards of the discipline. To this end, NABT will not support efforts to include in the science classroom materials or theories derived outside of the scientific processes. Nonscientific notions such as geocentricism, flat earth, creationism, young earth, astrology, psychic healing and vitalistic theory, therefore, cannot legitimately be taught, promoted, or condoned as science in the classroom.

Revision adopted by the Board of NABT, 3/15/95

NATIONAL ASSOCIATION OF BIOLOGY TEACHERS: *The Teaching of Evolution*

As stated in *The American Biology Teacher* (1973) by the eminent scientist Theodosius Dobzhansky, 'nothing in biology makes sense except in the light of evolution." This often quoted assertion accurately illuminates the central, unifying role of evolution in nature, and therefore in biology. Teaching biology in an effective and scientifically honest manner requires classroom discussions and laboratory experiences on evolution.

Modern biologists constantly study, ponder, and deliberate the patterns, mechanisms and pace of evolution, but they do not debate evolution's occurrence. The fossil record and the diversity of extant organisms, combined with modern techniques of molecular biology, taxonomy and geology, provide exhaustive examples and powerful evidence for genetic variation, natural selection, speciation, extinction and other well-established components of current evolutionary theory. Scientific deliberations and modifications of these components clearly demonstrate the vitality and scientific integrity of evolution and the theory that explains it.

This same examination, pondering and possible revision has firmly established evolution as an important natural process explained by valid scientific principles, and clearly differentiates and separates science from various kinds of non-scientific ways of knowing, including those with a supernatural basis such as creationism. Whether called "creation science, scientific creationism, intelligent-design theory, young-earth theory" or some other synonym, creation beliefs have no place in the science classroom. Explanations employing non-naturalistic or supernatural events, whether explicit reference is made to a supernatural being or not, are outside the realm of science and are not part of a valid science curriculum. Evolutionary theory, indeed all of science, is necessarily silent on religion and neither refutes or supports the existence of a deity or deities.

Accordingly, the National Association of Biology Teachers, an organization of science teachers, endorses the following tenets of science, evolution and biology education:

- The diversity of life on earth is the outcome of evolution: an unsupervised, impersonal, unpredictable and natural process of temporal descent with genetic modification that is affected by natural selection, chance, historical contingencies and changing environments.

- Evolutionary theory is significant in biology, among other reasons, for its unifying properties and predictive features, the clear empirical testability of its integral models and the richness of new scientific research it fosters.

- The fossil record, which includes abundant transitional forms in diverse taxonomic groups, establishes extensive and comprehensive evidence for organic evolution.

- Natural selection, the primary mechanism for evolutionary changes, can be demonstrated with numerous, convincing examples, both extant and extinct.

- Natural selection — a differential, greater survival and reproduction of some genetic variants within a population under an existing environmental state — has no specific direction or goal, including survival of a species.

- Adaptations do not always provide an obvious selective advantage. Furthermore, there is no indication that adaptations — molecular to organismal — must be perfect; adaptations providing a selective advantage must simply be good enough for survival and increased reproductive fitness.

- The model of punctuated equilibrium provides another account of the tempo of speciation in the fossil record of many lineages: it does not refute or overturn evolutionary theory, but instead adds to its scientific richness.

- Evolution does not violate the second law of thermodynamics; producing order from disorder is possible with the addition of energy, such as from the sun.

- Although comprehending deep time is difficult, the earth is about 4.5 billion years old. Homo sapiens has occupied only a minuscule moment of that immense duration of time.

- When compared with earlier periods, the Cambrian explosion evident in the fossil record reflects at least three phenomena: the evolution of animals with readily fossilized, hard-body parts; a

Cambrian environment (sedimentary rock) more conducive to preserving fossils; and the evolution from pre-Cambrian forms of an increased diversity of body patterns in animals.

- Radiometric and other dating techniques, when used properly, are highly accurate means of establishing dates in the history of the planet and in the history of life.

- In science, a theory is not a guess or an approximation but an extensive explanation developed from well-documented, repro-ducible sets of experimentally-derived data and from repeated observations of natural processes.

- The models and the subsequent outcomes of a scientific theory are not decided in advance, but can be, and often are, modified and improved as new empirical evidence is uncovered. Thus science is a constantly self-correcting endeavor to understand nature and natural phenomena.

- Science is not teleological: the accepted processes do not start with a conclusion then refuse to change it, or acknowledge as valid only those data that support an unyielding conclusion. Science does not base theories on an untestable collection of dogmatic proposals. Instead, the processes of science are characterized by asking ques-tions, proposing hypotheses and designing empirical models and conceptual frameworks for research about natural events.

- Providing a rational, coherent and scientific account of the taxo-nomic history and diversity of organisms requires inclusion of the mechanisms and principles of evolution.

- Similarly, effective teaching of cellular and molecular biology requires inclusion of evolution.

- Specific textbook chapters on evolution should be included in biology curricula, and evolution should be a recurrent theme throughout biology textbooks and courses.

- Students can maintain their religious beliefs and learn the scientific foundations of evolution.

- Teachers should respect diverse beliefs, but contrasting science with religion, such as belief in creationism, is not a role of science. Science teachers can, and often do, hold devout religious beliefs, accept evolution as a valid scientific theory and teach the theory's mechanisms and principles.

- Science and religion differ in significant ways that make it inappropriate to teach any of the different religious beliefs in the science classroom.

Opposition to teaching evolution reflects confusion about the nature and processes of science. Teachers can, and should, stand firm and teach good science with the acknowledged support of the courts. In *Epperson v. Arkansas* (1968), the U.S. Supreme Court struck down a 1928 Arkansas law prohibiting the teaching of evolution in state schools. In *McLean v. Arkansas* (1982), the federal district court invalidated a state statute requiring equal classroom time for evolution and creationism.

Edwards v. Aguillard (1987) led to another Supreme Court ruling against so-called "balanced treatment" of creation science and evolution in public schools. In this landmark case, the Court called the Louisiana equal-time statute "facially invalid as violative of the Establishment Clause of the First Amendment, because it lacks a clear secular purpose." This decision — "the Edwards restriction" — is now the controlling legal position on attempts to mandate the teaching of creationism: the nation's highest court has said that such mandates are unconstitutional. Subsequent district court decisions in Illinois and California have applied "the Edwards restriction" to teachers who advocate creation science, and to the right of a district to prohibit an individual teacher from promoting creation science in the classroom. Courts have thus restricted school districts from requiring creation science in the science curriculum and have restricted individual instructors from teaching it. All teachers and administrators should be mindful of these court cases, remembering the law, science and NABT support them as they appropriately include the teaching of evolution in the science curriculum.

REFERENCES AND SUGGESTED READING

Clough, M. (1994) Diminish Students' Resistance to Biological Evolution. *The Am. Biol. Teacher*, 56 409-415

Futuyma, D. (1986). *Evolutionary Biology*, 2 ed. Sunderland, MA: Sinauer Assoc., Inc.

Gillis, A. (1994) Keeping Creationism Out of the Classroom. *Bioscience*, 44, 650-656.

Gould, S. (1977). *Ever Since Darwin: Reflections in Natural History*. NY: W. W. Norton & Co.

Mayr, E. (1991) *One Long Argument: Charles Darwin and the genesis of modern evolutionary thought*. Cambridge, MA: Harvard Univ. Press.

McComas, W. ed. (1994). *Investigating Evolutionary Biology in the Laboratory*, Reston, VA: NABT.

Moore, J. (1993). *Science as a Way of Knowing – The Foundations of Modern Biology*. Cambridge, MA: Harvard Univ. Press.

National Center for Science Education. P.O. Box 9477, Berkeley, CA 94709. Numerous publications such as *Facts, Faith and Fairness – Scientific Creationism Clouds Scientific Literacy* by S. Walsh and T. Demere.

Numbers, R. (1992). *The Creationists: The Evolution of Scientific Creationism*. Berkeley, CA: Univ. Calif. Press.

Weiner, J. (1994). *Beak of the Finch – A Story of Evolution in Our Time*. NY: Alfred A. Knopf.

Adopted by the Board of the NABT, 15 March, 1995

NATIONAL EDUCATION ASSOCIATION
Statement in Support of the Teaching of Evolution

The National Education Association (NEA) was founded in 1857, two years before Charles Darwin published *The Origin of Species*. Although these two events remain unrelated, Darwinism and evolution came to play prominent roles over the next fifty years in the science curricula in our nation's public schools.

But like so many scientific theories that challenge established orthodoxy, evolution is still being contested. The issue of evolution versus creationism, unresolved by the weight of case law, is still the subject of debate.

NEA's position in this debate has been firm. Most recently, our 1982 Representative Assembly made clear that NEA opposes all efforts to alter the science curricula in any way that would place the teaching of scientific creationism on an equal footing with the teaching of evolution.

While the National Education Association believes that educational materials should accurately portray the influence of religion in our nation and throughout the world, we also believe that for American education to flourish, religious dogma must neither guide nor hamper the pursuit of knowledge by students and teachers in our public schools.

1994

Oklahoma Science Teachers Association

The scientific content of science courses should be determined by scientists and science educators and not by political directives. In particular, science teachers should not be required to teach, as science, ideas, models and theories that are clearly *extra-scientific*. An extra-scientific hypothesis, as such, might legitimately be discussed in a science class when examination of its logical construction and criteria for acceptance would illuminate the corresponding features of a scientific hypothesis and scientific method. Any requirement for equal time for such hypotheses is not justifiable.

Scientific hypotheses have a number of distinguishing properties, the foremost of which is that one should be able to deduce, from the basic postulates, logical consequences that can be tested against observation. Attention should be paid to the possible kinds of evidence that would falsify the hypothesis, rather than just the evidence that might confirm it. Other properties include:

1. The hypothesis should have more general consequences than those observations which initially suggested it. Thus it should be independently testable and not ad hoc.

2. It should be fruitful, suggesting new lines of research to pursue, raise new questions to by investigated by future research.

3. It should be logically consistent.

4. It should be consistent with the general scientific philosophy that the observed phenomena of the universe are real and that nature is consistent and understandable, that is, describable and explainable in terms of laws and theories.

Hypotheses that postulate miracles or supernatural events are falsified scientifically because they explicitly admit they cannot explain phenomena within their sphere of application. Furthermore, they are extra-scientific and non-explanatory because those phenomena are declared to be beyond human understanding. Thus they can not be considered alternate explanations to any scientific hypothesis because, by their very nature, they are anti-explanatory, seeking only to establish and perpetuate a mystery or mysteries.

All such hypotheses, models and theories that claim to be scientific should be required to meet the same criteria as do those hypotheses commonly considered to be scientific by the scientific community at large.

Adopted October 15, 1981
(later adopted by the Oklahoma Academy of Sciences)

SCIENCE MUSEUM
OF MINNESOTA*

As an institution whose mission is to invite learners of all ages and backgrounds to experience the world through science, this museum must be consistent in the meaning given to "science". By definition, science is knowledge derived from observation, study, and experimentation. Science encompasses a wide variety of disciplines. Each discipline has a characteristic focus but all are united by use of the scientific method, and all are affected by censorship.

There are few areas of life in which one will not encounter a degree of censorship. But since each of the various disciplines of science is bound indissolubly to the others, if one topic is omitted through censorship, the ability to study any of them is inhibited. While the study of biology focuses on organisms, it is forever dependent on chemistry, chemistry on physics, physics on mathematics, and so on. All scientific disciplines are united in demonstrating the evolution of life on this planet.

In every area of scientific research and education, one strives to remain consistent in vocabulary. "Theory" is just one of many words that has a different meaning in the world of science from the meaning it has in daily life. In daily life, one definition of "theory" is, "a mere guess at something." However, a *scientific* theory reflects an enormous amount of study that has gone into accounting for some natural phenomenon, and in science the word "theory" is not used lightly. As for the theory of evolution, it is widely accepted within the scientific community that evolution itself is fact. It is theory about the *mechanisms* of evolution that continues to be refined.

The Science Museum of Minnesota is currently undergoing the process of developing internal policies concerning discussion of evolution. Appropriate information is provided for staff in order to educate them and allow them to conduct informed discussions on the topic. In instances where creationists visit the museum, they are not discouraged providing they are not disruptive to the staff or other visitors. Leafleting of any kind is not allowed within this institution. Following is a list of critical issues scientific institutions must decide upon when striving to fulfill their missions in research, practice and education.

THE AGE OF THE EARTH

In order to carry on consistent conversations on a variety of topics, scientists must agree on the age of the earth. An educational institution cannot seriously discuss topics such as geology, biodiversity, human biology, embryology, ecology, paleontology, anthropology, and so forth, without first establishing a timeline of events. Since creationist doctrine provides a myriad of options as to the age of the earth, it does not lend itself to this process and therefore cannot be used. Based on current research, scientists generally agree that the age of the earth is approximately 4.5 billion years. An institution of scientists and science educators are obliged to use this date until further study finds otherwise.

EDUCATIONAL OBJECTIVES

Being true to educational objectives requires honesty. If science educators are to compare the enormous variety of life forms which have inhabited the planet, they must account for both the similarities and differences in those animals. Evolution is the framework within which these topics can be discussed. In addition, *evolution* applies to all life forms, not just some. It is the scientific institution's responsibility to the public not to negate pertinent information on the basis that it may not be acceptable to all.

SPEAKING FREELY ABOUT SCIENCE

If an institution is bound by censorship of topics fundamental to its work, it is of little use in either the educational or the scientific arena. If instead, the bounds of censorship are lifted, the quality of information that can be provided to the public becomes unlimited. Evolution is fundamental in the scientific discussion of life on earth and of the earth itself.

PROVIDING CLEAR GUIDELINES TO STAFF

An institution owes its staff clear guidelines on controversial topics so that they may convey the institution's position. However, it must also respect the rights of its staff to live by whatever ideology or doctrine they choose. An institutional policy statement does not prevent controversy, but since front line staff are the ones most likely to encounter difficulties, institutional support will aid in their handling of situations that arise. Staff are not required to agree with evolution, but they are expected to be able to provide direct answers to the public as to why the *institution* supports evolution. Staff should not be expected to defend their personal beliefs to visitors.

BEING HONEST WITH VISITORS

An institution has a responsibility to its visitors to provide a simple, concise and unbiased explanation as to why it accepts the evidence for evolution. While some visitors may disagree, they will not be led astray or told untruths. In an institution of science, visitors should expect to see all aspects of science within that institution's programs. The institution should be free to discuss science without regret or apology.

1995

*Editor's note: Official position statements of the Science Museum of Minnesota are not public documents; other, similar institutions should direct requests for further information to Patty Forber, Manager, Paleontology Science Hall, Science Museum of Minnesota, 30 East 10th Street, Saint Paul, MN 55101. We are grateful to the Museum for submitting this essay specifically for publication in *Voices for Evolution*.

University of California, Academic Council of the Academic Senate

It is our understanding that within the next few months the California State Board of Education will be approving many science textbooks for use in California public schools, grades K through 8. The text of the "Science Framework for California Schools" prepared in 1969, suggests that one criterion for the Board's approval of a text may be the extent to which, in the discussion of the origins of life, a "special theory of creation" is treated as a scientific theory in a manner parallel to an account of evolution. We believe that a description of special creation as a scientific theory is a gross misunderstanding of the nature of scientific inquiry.

To provide the basis of a scientific theory, an hypothesis must make testable predictions. Our ideas of biological evolution are continually being tested in the process of an enormous amount of investigation by thousands of professional biological scientists throughout the world. As in all sciences, there are many facets of the evolution picture that are not yet thoroughly understood, and researchers at the frontier of knowledge, often in disagreement with each other concerning details, continually revise their thinking. Thus, evolutionary theory itself has evolved considerably since the time of Darwin. But virtually all biological scientists are agreed on the broad features of the theory of evolution of life forms, the evidence for which is completely overwhelming.

The issue is not whether the concept of a relatively sudden special creation is true or valid, but rather that its origin lies in philosophical thought and religious beliefs, not in scientific investigation. Partly because of the wide diversity of religious opinions regarding creation, and especially because our traditional adherence to the First Amendment of the United States Constitution requires the separation of religious instruction from State-supported schools, we believe that the teaching of special creation should be avoided entirely in California public schools; certainly, it should not be presented in textbooks as a scientific theory.

We join the National Academy of Sciences, the American Association for the Advancement of Science, and other learned societies in urging the State Board of Education to reject inclusion of an account of special creation in State-approved science textbooks.

Statement approved by The Academic Council of The University of California Academic Senate on October 27, 1972, for transmittal to the California State Board of Education for their meeting on November 9, 1972

THE UNIVERSITY OF
QUEENSLAND (AUSTRALIA)
BOARD OF THE FACULTY OF SCIENCE*

Ifully support the decision of the Board of Secondary School Studies and its Science Advisory Committee to include the teaching of evolution as a component of the core syllabus for Senior Biology, and the decision not to include "Creation Science" as a compulsory component of Senior biology. Indeed "Creation Science" as it is espoused by its supporters has no place in the syllabus of any science subject....

---

*On May 6, 1984, the Board of the Faculty of Science at the University of Queensland resolved to endorse their Dean's letter to the Minister for Education, supporting teaching evolution in the secondary schools. The above statement is excerpted from that letter, recorded as a resolution in the minutes of the Board meeting.

UTAH SCIENCE TEACHERS ASSOCIATION

W*hereas*, the science teachers of the State of Utah are being subjected to increasing pressure to teach non-science material in their science classrooms, and,

Whereas, the Utah Science Teachers Association supports the wisdom and constitutionality of the separation of church and state,

The Utah Science Teachers Association hereby affirms that the science teachers of the State of Utah should[1]:

1. Teach science and related disciplines (technology, societal implications of science and technology, etc.) in their science classrooms, and not teach religion as science.

2. Teach students that science is a dynamic, self-correcting discipline based on empirical data and reasonable analyses thereof.

3. Teach the theory of evolution as the major organizing theory in the discipline of the biological and geological sciences.

4. Teach the students to distinguish between various types of evidence; to distinguish "fact," "theory," "hypothesis," "inference," etc.; and to recognize that in its strict sense, "theory" (as a generalization organizing massive amounts of diverse and repeatedly-tested data), is the most useful statement that life science can make.

5. Help students understand that accepting the theory of evolution by natural selection, and other biological phenomena, is not equating science with atheism and that the theory of evolution by natural selection does not rule out the possibility of the involvement of a divine Creator.

6. Help students understand that accepting the theory of evolution by natural selection need not compromise their religious beliefs, whatever their religion may be, since science and religion are based on separate premises and use different methodologies.

7. Help students understand that creationism, as taught by prominent creationist organizations of the day, is pseudoscience and not science.

8. Help students understand that religion is a belief system based on faith and religious experience, and that religious principles can still be followed without conflict while accepting the premises and methodology of science.

9. Help students understand that both science and religion, as two among several human endeavors, have strengths and limits in pursuing human knowledge and action; that neither alone is a sufficient guide for either individual or group conduct. It has never been an endeavor of science, nor is it appropriate for individual scientists, to falsely apply the methodology of science to undermine matters of religious faith.

Resolution adopted January 27, 1990

[1] *Statements 1 through 4 refer to actual classroom teaching recommendations. Statements 5 though 9 refer to suggestions teachers may want to consider in helping students outside of class.*

PART FIVE

Civil Liberties Organizations

AFRICAN AMERICANS
FOR HUMANISM (AAH)

I n recent years, religious fundamentalists have increased their efforts to teach Creationism in the public schools as an alternative to the theory of evolution. But though Creationism is pushed by religious adherents, the real conflict is not merely between religion and science, but between science and pseudoscience. Creationism does not qualify as a scientific theory because it begins with a conclusion (i.e., God created the universe) and seeks to support it, while scientific theories are prone to change (and may even be dismissed) in the light of new evidence. Creationism may or may not be good religion, but it is not good science, and should have no place in the public schools.

Many Creationists assert that evolution is used to further racism. But the scientific evidence has not led to racist conclusions in reputable scientific circles. On the contrary, human diversity is regarded as a product of genetic processes and natural selection, and "races" are always changing, often as a result of intermarriage among various peoples.

Conversely, many Creationists have propagated the racist "myth of Ham," or the belief that the "colored" peoples (who are supposedly descended from the eponymous Ham, the son of Noah) are cursed by God with servitude to whites. (Not surprisingly, such thinking spawned counter-myths among some black groups, such as the Nation of Islam, whose members have asserted that whites are a race of devils.) Evolution, far from supporting such notions, helps to dispel them.

Moreover, AAH is concerned that Blacks and other minorities are woefully underrepresented in the sciences. It will become increasingly difficult to attract and retain minority students to the sciences if they are constantly bombarded with pseudoscientific misinformation and unscientific methods of investigation. For these reasons, AAH opposes the introduction of Creationism into all science curricula of the U.S. public schools.

1994

AMERICAN CIVIL LIBERTIES UNION:
Position Statement on Creationism and Public Schools

For seventy-five years, the American Civil Liberties Union has been dedicated to upholding First Amendment protections of civil liberties. Consistent with the requirements of the Establishment Clause, the ACLU policy on religion in public schools states that "...any program of religious indoctrination — direct or indirect — in the public schools or by use of public resources is a violation of the constitutional principle of separation of church and state and must be opposed...." In 1980, the Board of Directors further clarified this policy by stating, "ACLU also opposes the inculcation of religious doctrines even if they are presented as alternatives to scientific theories." "Creation science" in all its guises, for example "abrupt appearance theory" or "intelligent design theory", is just such religious doctrine.

Among the problems "creation-science" creates in the academic environment is the foreclosure of scientific inquiry. The unifying principle of "creationism" is not the law of nature, but divinity. A divine explanation of natural data is not subject to experiment, it cannot be proved untrue, it cannot be disputed by any human means. Creationism necessarily rests on the unobservable; it can exist only in the ambiance of faith. Faith — belief that does not rest on logic or on evidence — has no role in scientific inquiry.

The constitutional defect of any law or policy requiring the teaching of creationism, or of "evidence against evolution," is not that it requires instruction about facts which coincide with a religious belief, but that it requires instruction in one religious belief as the unifying explanation of facts. This unifying concept is not a secular topic such as biology, chemistry, art, phonics, or literature which is familiar to the elementary and secondary school curricula. Instead, teachers are required to identify, organize, or teach facts and inferences supporting a specific belief — "special creation". To require public schools to marshal "evidences" and "inferences" in service of one religious belief, or to impose an embargo on a scientific theory that Fundamentalists dislike, is not to use religious works "for the teaching of secular subjects," (*Abington School Dist. v. Schempp*), but to place "the power, prestige and financial support of government...behind a particular religious belief" (*Engel v. Vitale*). The year-by-year, school-by-school, and teacher-by-teacher decision-making on whether and how to imbue "creationism" into the sciences and humanities promises continuing anguish in the educational community and assures inordinate involvement of religious groups in the affairs of government.

In our society, government is not permitted to instruct a child in religion, because it is not the government's job to promote a religious form of truth. No provision of the Constitution so firmly assures the essential freedom of the individual as does the Establishment Clause. The provision recognizes that choices about the ultimate meaning of life must be made in the private recesses of the conscience and not in the earthly controversies of political power. Were every person in this country of the same faith, the Establishment Clause would serve as a powerful expression that humans must decide their relationship to God, not at the bidding of the state, but at the calling of the soul. That we are a nation of many religions does not alter this basic function of the Clause; it only enhances the need for vigilance against state manipulation of belief.

Vigilance requires firm and consistent opposition to every effort to use the nation's schools to teach any biblical text, including Genesis, as literal truth, either directly or disguised as "alternative" science. To reject creationism as science is to defend the most basic principles of academic integrity and religious liberty.

1994

AMERICAN HUMANIST ASSOCIATION
A Statement Affirming Evolution as a Principle of Science

For many years it has been well established scientifically that all known forms of life, including human beings, have developed by a lengthy process of evolution. It is also verifiable today that very primitive forms of life, ancestral to all living forms, came into being thousands of millions of years ago. They constituted the trunk of a "tree of life" that, in growing, branched more and more; that is, some of the later descendants of these earliest living things, in growing more complex, became ever more diverse and increasingly different from one another. Humans and other highly organized types of today constitute the present twig-ends of that tree. The human twig and that of the apes sprang from the same apelike progenitor branch.

Scientists consider that none of their principles, no matter how seemingly firmly established — and no ordinary "facts" of direct observation either — are absolute certainties. Some possibility of human error, even if very slight, always exists. Scientists welcome the challenge of further testing of any view whatever. They use such terms as firmly established only for conclusions, founded on rigorous evidence, that have continued to withstand searching criticism.

The principle of biological evolution, as just stated, meets these criteria exceptionally well. It rests upon a multitude of discoveries of very different kinds that concur and complement one another. It is therefore accepted into humanity's general body of knowledge by scientists and other reasonable persons who have familiarized themselves with the evidence.

In recent years, the evidence for the principle of evolution has continued to accumulate. This has resulted in a firm understanding of biological evolution, including the further confirmation of the principle of natural selection and adaptation that Darwin and Wallace over a century ago showed to be an essential part of the process of biological evolution.

There are no alternative theories to the principle of evolution, with its "tree of life" pattern, that any competent biologist of today takes seriously. Moreover, the principle is so important for an understanding of the world we live in and of ourselves that the public in general, including students taking biology in school, should be made aware of it, and of the fact that it is firmly established in the view of the modern scientific community.

Creationism is not scientific; it is a purely religious view held by some religious sects and persons and strongly opposed by other religious sects and persons. Evolution is the only presently known strictly scientific and nonreligious explanation for the existence and diversity of living organisms. It is therefore the only view that should be expounded in public-school courses on science, which are distinct from those on religion.

We, the undersigned, call upon all local school boards, manufacturers of textbooks and teaching materials, elementary and secondary teachers of biological science, concerned citizens, and educational agencies to do the following:

- Resist and oppose measures currently before several state legislatures that would require that creationist views of origins be given equal treatment and emphasis in public-school biology classes and text materials.

- Reject the concept, currently being put forth by certain religious and creationist pressure groups, that alleges that evolution is itself a tenet of a religion of "secular humanism," and as such is unsuitable for inclusion in the public-school science curriculum.

- Give vigorous support and aid to those classroom teachers who present the subject matter of evolution fairly and who often encounter community opposition.

Composed by Bette Chambers, Isaac Asimov, Hudson Hoagland, Chauncy D. Leake, Linus Pauling, and George Gaylord Simpson; published over the signatures of 163 scientists, theologians, philosophers, and others in The Humanist, *37(1):4-6 (Jan/Feb 1977).*

AMERICANS FOR RELIGIOUS LIBERTY

A free and secular democratic state values education in science. It recognizes that a strong country needs citizens who are trained in the methods of science and makes it available through public institutions. Since it protects the integrity of science and free inquiry it refuses to allow public school classrooms to be used for religious indoctrination. It especially defends the integrity of modern biology. The evolution of life is science. It is more than speculation. It is an established truth, which over one hundred years of biological research has confirmed.

Approved by the Board of Directors, 1982

AMERICANS UNITED FOR
SEPARATION OF CHURCH AND STATE

n recent years, a great deal of conflict has erupted over the issue of religion in public education. Although some individuals and organizations have worked to interject sectarian dogma into the schools, the Supreme Court has repeatedly ruled that public education must remain neutral on religious matters.

One area of especially sharp conflict has been creationism. While all religious denominations espouse a particular theology regarding the origins of the universe and humankind, these theological beliefs vary widely among faith groups. "Creationism" as a term commonly used by Christian fundamentalists in this country refers specifically to the belief that the creation story found in Genesis 1 and 2 is literally true and that the universe and humankind were created by God 6,000 years ago. This view, which is at odds with modern scientific understanding, is not shared by all American Christians.

As such, the teaching of creationism as science in the public schools would promote a particular religious viewpoint and would discount the theologies of other faith groups, thus amounting to an establishment of religion and a violation of the First Amendment.

The Supreme Court has dealt with the issue twice. The Court ruled that public schools may not forbid the teaching of evolution just because some religious groups find it offensive (*Epperson v. Arkansas*, 1968) and that the teaching of creationism as science in public schools violates church-state separation since it is a theological concept (*Edwards v. Aguillard*, 1987).

Ideas concerning the origins of humans and the universe that are based on religion are appropriate when used within the context of religious education, such as sabbath schools and private church school instruction. These ideas are not appropriate for use in public schools, where students of many different religious faiths gather. Public school curricula — including science classes — must be kept free of sectarian dogma.

Public school educators and administrators should resist pressures to introduce creationism into science classes. While creationism could be discussed objectively in comparative religion courses or classes on the history of science, it has no place as a viable theory in science classes because it amounts to the introduction of sectarian dogma into the curriculum and violates the separation of church and state.

1994

COUNCIL FOR DEMOCRATIC AND SECULAR HUMANISM (CODESH)

Concerning the origin and historical diversity of life on earth, secular humanists accept the fact of evolution as the essential framework of modern biology. Physico-chemical development paved the way for the origin of life about four billion years ago. Subsequent organic evolution is now documented by empirical evidence from geology, paleontology, biogeography, anthropology and genetics as well as comparative studies in taxonomy, biochemistry, embryology, anatomy and physiology. The ages of rock strata, with their fossils and artifacts in the geological column, are determined by radiometric dating techniques. Grounded in science and reason, evolution has descriptive and explanatory powers free from supernatural claims and dogmatic religious beliefs. Concerning models, mechanisms and interpretations, the present Neo-darwinian synthesis in biological evolution is always subject to modification and expansion in light of new discoveries in science and widening perspectives in philosophy.

Defending the constitutional separation of church and state, secular humanists deplore the efforts of biblical fundamentalists or so-called scientific creationists to invade science classrooms and pressure textbook publishers with their religious myth and political agenda. We reject the teaching of religious fundamentalism as a viable alternative to organic evolution in science texts and biology classes. In fact, all religious beliefs and practices have evolved throughout human socio-cultural development. Clearly, a strict and literal interpretation of Genesis is merely a religious account for the origin of life that is not subject to testing by evidence, experience and experimentation. Consequently, biblical creationism is an ongoing and serious threat to science education, responsible research, critical thought and free inquiry. Authority and revelation are not reliable substitutes for the scientific method and logical procedure. In short, rigorous scrutiny shows evolutionary science and scriptural literalism, with its appeals to miraculous causes, to be opposing explanations for the appearance of all life forms on this planet.

Furthermore, secular humanists boldly accept the far-reaching consequences of evolution and extinction for understanding and appreciating the place our species occupies within earth history and this dynamic universe. The human animal is a product of, dependent upon, and totally within organic evolution. Comparative DNA studies show that humankind shares a common ancestry with the three great apes (orangutan, chimpanzee and gorilla). Fossil hominid evidence recently found in central East Africa documents the emergence of our species over the past four million years. No doubt, future discoveries will shed additional light on the origin and history of humankind from ape-like ancestors.

Religious beliefs in a personal god, human immortality, and a divine destiny for our species are inadmissible as scientific statements. And questions concerning metaphysics, epistemology, ethics and values are best answered in terms of science, reason and human experience within a humanist framework and a naturalist worldview.

*Drafted for CODESH by H. James Birx, Ph.D. Executive Director for the
Alliance of Secular Humanist Societies (ASHS), October, 1994*

FREEDOM FROM
RELIGION FOUNDATION

E volution is a fact, and schools should teach facts.

The phrase "theory of evolution" does not suggest uncertainty about the fact of evolution any more than the phrase "music theory" questions the existence of music. A theory is a framework by which a known process is understood.

The prevailing theory of biological evolution is Darwin's idea of the hereditary transmission of slight variations through successive generations. Some variations are naturally "selected" due to adaptiveness. Biology makes no sense without recognizing the fact that all species of plants and animals (including humans) have developed from earlier forms. Natural selection has withstood more than a century of rigorous scientific testing.

Creationism, a religious belief, has withstood no testing. Whereas scientists will tell you exactly what would falsify evolution (for example, routinely discovering horse skeletons mixed in with trilobite fossils in the Cambrian strata), creationists never volunteer what set of circumstances, if true, would count against their idea that all species emerged at one time. Since creationism is not assailable, not vulnerable to experiment, it is not science.

The bulk of creationist literature consists of attacks against evolution, pretending that the eradication of the idea of evolution would cause creationism to win by default. The only "evidence" creationists present is the story in Genesis, or other religious texts, that must be accepted by faith, not by rational principles of verification.

Creationism can be discussed in the context of comparative religion, philosophy, politics, or culture. It should not be taught in the science classroom.

Many religious people welcome the fact of evolution, just as they accept the theory of relativity with no threat to their faith. They see evolution as one of the tools their God used in creation.

All human beings, religious or not, should feel enriched by discovering our place in nature and the interconnectedness of all living things. The understanding of evolution by natural selection is wonderfully enlightening to science. It should be loudly and proudly taught.

HUMANIST ASSOCIATION OF CANADA

E volution is the basis of modern biology. A student cannot possibly understand any of the life sciences without understanding the process of evolution that is the foundation of these sciences. It is the unifying web that links them together. Without evolution, biology is only a series of disconnected facts. With evolution, comes a comprehension of adaptation to local ecologies, the relationships among species, and the relationships among plants and animals and environments.

The physical sciences also depend on an accurate knowledge of the origins of the universe, radioactive decay, the age of the earth, chemical reactions and many physical relationships that change uniformly over time.

All these facts complement and reinforce each other. To comprehend the age of the universe, students must learn about the speed of light, what a red shift is, and something about spectroscopy. To understand the age of the earth, they must understand sedimentation, fossilization, and radioactive decay. To understand human origins they must understand genetic drift, carbon dating, archeology and linguistic change. Then they can begin to appreciate how these different systems reinforce each other and mutually confirm many diverse facts. They will then appreciate how a scientific hypothesis must fit in with all the observed and confirmed facts of the universe. Gravitation, relativity, radioactive decay, molecular genetics; one odd observation or a new theory cannot overturn these well-established truths. For instance, Einstein's relativity did not change the observations Newton had made. Newton's laws of gravity and motion still work correctly, unless one is dealing with objects moving nearly at the speed of light. Students must learn that a new theory must still fit in with the old facts, while explaining observations that were previously unexplained.

Evolution is an area where great strides are being made yearly, especially in the area of human origins. Students should learn how science changes because of new information.

Students should learn how science progresses through the accumulation of facts, through testing theories to see if they explain the facts, or if any facts disprove the theory.

Science education must avoid implying that science is a received and unchangeable set of dogmas for students to memorize, but must also teach those facts that are true.

So-called creation science is merely misnamed religious propaganda. There is nothing of science in it beyond the name. Creationists do not accept any evidence that goes against their case and will not listen to any arguments that don't go their way. Science is interested in the truth, and seeks out tests that might disprove a new theory. Creationists do not recognize anything that contradicts their beliefs. They have simply decided in advance what the conclusion must be, and head toward it without consideration of any contrary evidence that may be in the path. Creationism is an attempt to sneak a narrow sectarian religious creed into public classrooms and impose it on everyone. It is based, despite denials, upon the Genesis texts of the Jewish and Christian bible. Only a few Christian denominations, and a few Jewish sects, demand literal interpretation of these myths. Creation science is a religious dogma. Most of all, it is false and it is inaccurate. There is no evidence to support the creationist argument.

There is no room in science for believing without evidence. The whole of humanity's scientific enterprise is based on testing every belief and relinquishing those that fail the tests. Any public school teacher or public school board that lets "creation science" into a classroom, has abandoned teaching and taken up preaching and should be stopped.

INSTITUTE FOR FIRST AMENDMENT STUDIES:
The Case for Evolution

A popular bumper sticker reads: "God says it, I believe it, that settles it." For most Christian fundamentalists, that statement neatly sums up their belief in Biblical inerrancy. They believe in creationism because the Bible says that God created everything in six days at some point less than 10,000 years ago.

"Creation scientists" take that viewpoint a step further. By faith they begin with belief in creationism — then they search for evidence to back that belief.

True scientists study the evidence, drawing their conclusions from that evidence. Science does not deal in "truths," but in models which have predictive values. Evolution is a truly scientific model; it is open to examination and challenge. Over the years scientists have modified their evolutionary viewpoints to fit the latest evidence. Because it is Bible-based, creationists never modify their hypothesis, or even admit it could be in error.

Creationism is clearly based upon religion. As such, teaching it in church, Sunday school, parochial school (or even in comparative religion classes in public school) is fine. However, because it is faith-based, teaching creationism as science in tax-supported public schools violates the separation between church and state.

_1994_

THE NATIONAL COMMITTEE
FOR PUBLIC EDUCATION
AND RELIGIOUS LIBERTY

The National Committee for Public Education and Religious Liberty (National PEARL) is a coalition of over fifty* grass-roots, civic, educational and religious groups committed to maintaining the First Amendment's guarantee of separation of church and state in our nation's public schools. National PEARL believes that maintenance of the wall of separation helps to assure a strong public education system and safeguards religious liberty. National PEARL is committed to keeping the nation's public schools a safe haven for the nation's children, free of religious indoctrination and discrimination.

National PEARL opposes teaching creationism, in lieu of or as a "companion" theory to, theories of scientific evolution in public schools. There are several versions of creationism; all share the common view that life, matter, and the universe were designed and created by a divine creator/supreme spiritual being. According to many creationists, all life developed relatively recently. Creationism cannot be taught without reference to the religious ideology from which it springs, namely the account of Genesis in the Bible. Consequently, National PEARL holds that creationism is a form of religious belief.

The teaching of creationism in a public school amounts to use of state-financed, state-run schools to indoctrinate children in a particular set of religious beliefs. This is best demonstrated by the fact that when creationists demand creationism be taught, they insist on the exclusion or denigration of legitimate science. For example, the Louisiana state legislature's consideration of legislation in 1981 that prohibited "discrimination" against teaching creationism but did not prohibit "discrimination" against teaching evolution.

AS A MATTER OF EDUCATION POLICY

A host of thorny educational issues arise from teaching creationism. These problems generate strife among teachers, between teachers and administrators, students and teachers, parents and the school, parents and students, and among students. If creationism were taught in the schools, it would foment religious strife over the following issues:

Who writes the curriculum? How could a religious curriculum be monitored objectively? Could an administrator require a teacher to teach creationism? If students attempted to opt out of the lesson, how would they be graded, much less treated? What if a teacher refuses to teach creationism?

Teaching creationism would mean that a teacher could answer a student's questions by reference to the book of Genesis or materials that are designed to support a theory of creation that is consistent with Genesis. Teaching creationism in lieu of science could also open a Pandora's box by requiring teachers to teach other religious or less-than-scientific views of other topics, on the theory that if the Biblical treatment of an issue is permitted, all other religious treatment of other scientific issues must have "equal access" to student's minds to avoid inter-religious strife. Conceivably, a Wicca theory of fire, or the Aryan Nation's or the Church of the Creator's theories that God did not create all people equal because some, by virtue of their race, are inferior, or other views like these would have to be permitted in science classes if creationism were permitted.

As a result, students would be presented with a dizzying array of religious doctrines but would not have the scientific training necessary to evaluate them or compete with other students. Preparing students to be well-informed and well educated is the cornerstone of the public school system, and concomitantly, of a functioning democracy.

This is not a case of abrogation of teachers' academic freedom. Proponents of creationism incorrectly appropriate the notion of academic freedom to argue for the right to teach their religious views. Proponents of creationism cannot equate academic freedom with their intent to indoctrinate students in a public school. The fact is, teachers' academic and religious freedom is undermined when they are forced to teach religious doctrines in science class.

Notably, no major union of teachers, including the National Education Association and the American Federation of Teachers, have ever characterized it in this manner. Most teachers are perfectly capable of simultaneously holding private, religious beliefs and teaching scientific evolution. In fact, teachers throughout the United States espouse the sentiment of the Louisiana Science Teachers Association, which stated in 1981 it considered creationism "to be outside the boundaries of bona fide science."

As a Matter of Law

Teaching creationism is impermissible as a matter of law, either in lieu of scientific evolution or as a "companion theory." In both contexts, it has continuously been found to violate the Establishment Clause of the First Amendment of the U.S. Constitution because it puts government-run schools in the position of establishing religion by using their power to teach children compelled to attend school.

Precisely because the state would use its power, in the form of publicly financed schools, to further a particular religious doctrine, teaching creationism violates the major precept of the Establishment Clause, namely that "neither [a state nor a federal government] can pass laws which aid one religion, aid all religions, or prefer one religion over another.' Everson v. Board of Education, 330 U.S. 1, 15 (1947). This kind of governmental support for private, religious belief and indoctrination goes against the philosophy of the Founding Fathers when they wrote the First Amendment. That such teachings are promulgated by legislative authorities, not educational experts, testifies to the reality that the real motivation and purpose is the advancement of a particular religious ideology.

Application of the most widely used legal test, known as *Lemon v. Kurtzman,* 403 U.S. 602 (1971), to the practice of teaching creationism in public schools has found it unconstitutional. See *Edwards v. Aguillard,* 482 U.S. 578 (1987). Under *Lemon,* if a practice has a) a religious purpose, b) the effect of advancing religion, or c) it causes or necessitates entanglement of church and state to administer it, the practice violates the Establishment Clause.

Under the "endorsement" test, which courts often use in lieu of or in conjunction with the Lemon test, a practice is judged according to how much the state is perceived as endorsing religion. Teaching creationism obviously violates this test because the power of the state is used to endorse a particular religious belief. Furthermore, there is no way to "mitigate" the state's endorsement of the religious message. As PEARL founder and noted constitutional scholar Leo Pfeffer reflected, "In respect to those pupils who do understand what the teachers are saying, teaching creationism as being only a theory would violate the First Amendment's ban on inhibiting religion. To teach pupils that the account of Moses splitting the sea or Jesus walking on it is only a theory could hardly be reconciled with the Amendment's ban on the inhibition of religion. The last thing in the world fundamentalist Christians want is for public schools to teach that God's creation of the world or His relationship to Jesus, or Moses' receipt of the Ten Commandments from Him, are only theories."

Under the "coercion" test, which courts often use in lieu of or in conjunction with the Lemon test, the teaching of creationism in public schools also violates the Establishment Clause. First, children are compelled to attend public school; they cannot "opt out" of science class and assume they will pass statewide, year-end tests. Consequently, forcing students to listen to creationist lectures would use students' captive status coercively. By the very nature of creationist theory, any student questioning or challenging the theory would be put in the position of questioning the religious belief system behind it, and risking the chance of invoking the disapproval of a teacher who espouses the creationist perspective.

For all the foregoing reasons — educational and constitutional — creationism should not be taught in the public schools.

March 1995

*American Association of School Administrators · American Association of University Women · American Civil Liberties Union · American Ethical Union, American Federation of Teachers · American Humanist Association · American Jewish Congress · Americans for Democratic Action · Americans for Religious Liberty · Americans United for Separation of Church & State (and Rochester Chapter) · Anti Defamation League · A. Philip Randolph Institute · Arizona Citizens Project · Association of Reform Rabbis of New York City & Vicinity · Baptist Joint Committee · Central Conference of American Rabbis · City Club of New York · Community Church of New York, Social Action Committee · Council of Churches of the City of New York · Council for Democratic and Secular Humanism · Council of Supervisors and Administrators · Episcopal Diocese of Long Island, Committee on Social Concerns & Peace · Episcopal Diocese of New York · Federation of Reconstructionist Congregations & Havurot · Freedom to Learn Network · Freethought Society of Greater Philadelphia · Humanist Society of Metropolitan New York, Inc. · Institute for First Amendment Studies · League for Industrial Democracy, NYC Chapter · Michigan Council About Parochiaid · Minnesota Civil Liberties Union · Monroe County PEARL · National Council of Jewish Women (& New York Section) · National Center for Science Education · National Education Association · National Emergency Civil Liberties Committee · National PTA · New York Jewish Labor Committee · New York Society for Ethical Culture · New York State Congress of Parents and Teachers · New York State Council of Churches · New York State United Teachers · Ohio PEARL · Public Education Association · Union of American Hebrew Congregations (& New York Federation of Reform Synagogues) · Unitarian-Universalist Association · United Community Centers, Inc. · United Federation of Teachers · United Synagogues of America, New York Metropolitan Region · Washington Area Secular Humanists · Women's American O.R.T. · · Women's City Club of NY, Inc. · · Workmen's Circle, NY Division

PEOPLE FOR THE AMERICAN WAY

P eople For the American Way emphatically opposes efforts to inject Creationism into the public school science curriculum. The Creationist "theory" is predicated upon religious beliefs; while it may be appropriate to discuss these beliefs in a comparative religion or social studies classroom, disguising them as science dilutes the quality and undermines the integrity of any science curriculum and, as the courts have recognized, violates the Establishment Clause of the First Amendment.

It is particularly troublesome that some science textbooks have inadequately covered or altogether omitted mention of evolution, for fear of inciting controversy. America's students deserve and require a high-quality science education, grounded in fact and free of sectarian influence.

1994